Neural Networks

Neural Networks

Phil Picton

palgrave

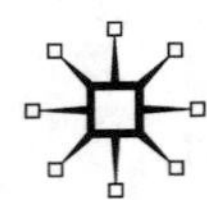

First published 2000 by
PALGRAVE
Houndmills, Basingstoke, Hampshire RG21 6XS and
175 Fifth Avenue, New York, N. Y. 10010
Companies and representatives throughout the world

PALGRAVE is the new global academic imprint of St. Martin's Press LLC Scholarly and Reference Division and Palgrave Publishers Ltd (formerly Macmillan Press Ltd).

ISBN 0–333–94899–8 hardcover

This book is printed on paper suitable for recycling and made from fully managed and sustained forest sources.

A catalogue record for this book is available from the British Library.

Library of Congress Cataloging-in-Publication Data has been applied for.

Typeset by Ian Kingston Editorial Services, Nottingham, UK

10 9 8 7 6 5 4 3 2 1
09 08 07 06 05 04 03 02 01 00

Printed in Great Britain by
Antony Rowe Ltd, Chippenham, Wilts

Contents

Preface

This book gives a very broad introduction to the subject of neural networks, and includes many of the dominant neural networks that are used today. It is ideal for anyone who wants to find out about neural networks, and is therefore written with the non-specialist in mind. A basic knowledge of a technical discipline, particularly electronics or computing, would be helpful, but is not essential.

Neural networks as part of artificial intelligence

For readers who are interested in the subject of artificial intelligence (AI) in general, it is worth starting off by putting neural networks in context. Needless to say, artificial intelligence is a very broad subject, and has branches in cognitive psychology, philosophy, mathematics, cybernetics and of course computer science. Broadly speaking, however, we can make a distinction by saying that some aspects of AI are concerned with understanding how the human brain or mind works, and attempts to model this using computers; in other words, modelling the whole brain as a single entity by trying to build a computer model that behaves in the same way. The main alternative is to try to look at the structure of the brain, which is built of billions of interconnected neurons, and to see if we can produce similar activity to the brain using networks of artificial neurons.

The first of these areas tends to concentrate on the rational part of the brain functions. Using techniques based around logic, usually in the form of a set of rules, quite sophisticated systems can be built. These include expert systems, which are able to outperform humans now in many applications such as medical diagnosis, game-playing (for example chess) and monitoring systems which don't get bored like humans! These have also been made extremely versatile with the introduction of fuzzy logic, which allows qualitative descriptions such as 'x is very cold' or 'y is quite small' to be incorporated into rules. These rules are often found in automated systems such as lift controllers or even driving underground railways in Japan.

The second technique is to use mathematical models of neurons to build artificial neural networks. By arranging the neurons into different configurations it is possible to model some of the brain's functions, such as being able to store information in memory, making associations between patterns, for example, how a particular scent may trigger an image of a particular place, and producing a reflex response so that a particular stimulus produces an appropriate reaction. Now, all of these could be done other ways, but the important feature is that a neural network learns to perform these functions itself rather than having them programmed in from the start. This means that many applications which have proved to be difficult in the past can now be implemented. Many of these, such as face recognition, for example, are difficult because we humans do it so easily but cannot rationalise about how we do it. If we cannot rationalise the problem we cannot represent the problem as a set of logical procedures; hence the difficulty in implementing this function in a rule-based system. Neural networks, on the other hand, can learn the associations between the image inputs of a face and the internal representation, such as the person's name.

Neural networks therefore have this ability to learn complex functions which have proven difficult for rule-based systems. However, neural networks do have one drawback. They also can't rationalise about what they do, so when a neural network has learnt to perform face recognition, it can't explain how it has done it. Much of the research that is still ongoing, therefore, is looking at hybrid systems which incorporate neural networks, so that they have the ability to learn, and rule-based systems so that they have the ability to explain what they are doing. When these two can work happily side-by-side, then we will have truly intelligent machines.

Recent developments in neural networks

Neural networks made a great come-back in the 1990s and are now generally accepted as a major tool in the development of intelligent systems. Their origins go way back to the 1940s, when the first mathematical model of a biological neuron was published by McCulloch and Pitts. Unfortunately, there was a period of about 20 years or so when research in neural networks effectively stopped.

It was during this period that I first became interested in neural networks, in the relatively obscure area of threshold logic, which is an attempt to replace the conventional building blocks of computers with something more like artificial neurons.

I was pleased but surprised when, in the mid 1980s, there was a resurgence of interest in neural networks, largely prompted by the publication of Rumelhart and McClelland's book, *Parallel Distributed Processors*. Suddenly, it seemed that everyone was interested in and talking about neural networks again.

It soon became apparent that during these lean years, now-eminent names such as Widrow, Kohonen and Grossberg had continued working on neural networks and developed their own versions. Problems such as the exclusive-or, which had originally contributed to the demise of neural networks in the 1960s, had been overcome using new learning techniques such as back-propagation. The result has been that researchers in the subject now have to familiarise themselves with a wider variety of networks, all with differences in architecture, learning strategies and weight updating methods.

Overview of the book

The book starts off in *Chapter 1* by describing how neural networks are able to perform functions which are almost impossible to implement any other way. Biologically inspired neural networks are introduced, as well as a unique alternative which behaves like a neural network but is actually built using conventional computer hardware.

Chapter 2 looks more closely at the model of a biological neuron, and describes how it can learn in a way that is analogous to way that the connections between biological neurons are strengthened or weakened chemically. This is placed in an historical context by looking at one of the earliest neural networks called the ADALINE.

Chapter 3 shows how the limitations of the ADALINE-type networks were overcome in the multi-layered perceptron. It was this discovery that brought neural network research back from its 20-year absence, and really kicked off much of the excitement that still persists today.

Chapter 4 returns to the idea of building neural networks using conventional computer hardware, and in particular looks at the development of the system know as the WISARD. Unlike most neural networks, this one can be easily built in hardware or software and operates very simply. It has all the properties of a neural network in that it can learn and perform pattern recognition, but in a way that is very different from biological or conventional artificial neural networks.

Chapter 5 looks at the properties of networks in which the outputs are fed back to the inputs. In general, what happens is that the networks have some sort of memory, and can store associations between input and output patterns. The main inspiration for these networks came from John Hopfield, so much of the chapter is taken up with the Hopfield network. More networks are examined and some of the limitations discussed.

Chapter 6 starts off by showing how statistics, and probability in particular, can be used to overcome these limitations of feedback networks. The Boltzmann machine is used as an example to show how a network with a random architecture, and neurons which only fire according to a probability, can seemingly pull order out of this chaos and behave better than the more conservative feedback networks. The chapter then introduces some feed-forward networks, such as the radial basis function network, which has become one of the main rivals of the multi-layered perceptron in popularity.

Chapter 7 then introduces the idea that networks can organise themselves and uses examples known as the adaptive resonance theorem, which has come out of cognitive psychology. In addition, the networks developed by Teuvo Kohonen in Finland are introduced, as they are regarded as one of the major developments in neural networks. His networks organise themselves in a similar way that it is believed neurons in the brain are arranged so that neurons that perform similar functions are close to each other.

Chapter 8 looks at the way that neural networks have been incorporated into control systems. One reason for including this chapter is that it isn't immediately obvious how you would design a control system using what is essentially a pattern recogniser. Examples of pattern recognition have been used throughout the book to illustrate the networks. This chapter shows how a traditional subject can be approached from a completely new angle because of neural networks.

Chapter 9 deals with threshold logic, which is an alternative approach to designing computers using neurons instead of logic gates. This work has been ongoing for many years and has produced some excellent theory on neural networks, which is often overlooked. The chapter introduces the concepts without delving too deeply.

Finally, *Chapter 10* looks at some of the attempts that have been made to implement a neural network in hardware, either electronically or optically. It is probably fair to say that, to date at least, software has always been the preferred implementation. This is mainly because the performance of computers is improving so rapidly that software stays one step ahead, since hardware takes a long time to develop. However, this chapter shows some of the basic building blocks that are used, and which may, one day, be the basis for an artificial brain!

CHAPTER 1

Introduction

CHAPTER OVERVIEW

This chapter gives a broad introduction to neural networks, particularly in the context of pattern recognition. It examines:

- what is meant by pattern classification
- how electronic memory can be used for pattern classification
- how a Boolean neural network learns and generalises
- how a McCulloch–Pitts neuron works
- how to find the output of a neuron

1.1 What is a neural network?

This is the first question that everybody asks. It can be answered more easily if the question is broken down into two parts.

Why is it called a neural network?
It is called a neural network because it is a network of interconnected elements. These elements were inspired from studies of biological nervous systems. In other words, neural networks are an attempt to create machines that work in a similar way to the human brain by building these machines using components that behave like biological neurons.

What does a neural network do?
The function of a neural network is to produce an output pattern when presented with an input pattern. This concept is rather abstract, so one of the operations that a neural network can be made to do – pattern classification – will be described in detail. Pattern classification is the process of sorting patterns into one group or another.

1.2 Pattern classification

As you are reading this sentence your brain is having to sort out the signals that it is receiving from your eyes so that it can identify the letters on the page and string them together into words, sentences, paragraphs and so on. The act of recognising the individual letters is pattern recognition, the symbols on the page being the patterns that need to be recognised.

Now, because this book has been printed, the letters are all in a particular typeface, so that all the letter 'a's, for example, are more or less the same. A machine could therefore be designed that could recognise the letter 'a' quite easily, since it would only need to recognise the one pattern.

What happens if we want to build a machine that can read handwritten characters? The problem is much more difficult because of the wide variation, even between examples of the same letter. It is therefore difficult to find a representative character for each of the letters, so a set of typical characters for each letter has to be constructed. Then, for example, all the different letter 'a's that are going to be read belong to the same set. Similarly, all the other letters belong to their own set, giving 26 sets in all. Each set of characters is called a class and we now have to build a machine that chooses which class an input pattern belongs to – a pattern classifier. This is still a hard problem, but it can be solved successfully because the input patterns do not have to match one of the characters in the class exactly. Instead, the machine has to be able to decide that the input pattern is more like the members of one class than any of the others, which is a simpler task.

The benefit of using pattern classes in this problem is that a handwritten character which is variable can be read into the system, and a printed character can be produced at the output. Other problems can also be solved using pattern classification. These problems usually involve the recognition of something that cannot be entirely described or predicted. Examples include recognising faces or fingerprints, identifying spoken words, and safety systems where you want to be warned if something unusual happens, which by definition cannot be predicted.

In essence then, a neural network can be used to solve or implement a problem if that problem can be reduced to pattern classification. In a large number of cases this is possible. The function of the neural network is to receive input patterns and produce a pattern on its output which is correct for that class.

It has been said that a neural network is a pattern classifier, so are all pattern classifiers neural networks? No – as will now be demonstrated.

The best way to explain this is to use an example. Imagine a situation where we want to make a machine that can read the numbers 0 to 9 from a document. The numbers could be typed or hand-drawn, and in a variety of different styles, sizes and possible orientations. So clearly any single numeral, 3 say, can have a large number of representations. Figure 1.1 shows some examples.

Any one of these inputs could appear, but we want the machine to respond in the same way each time, by indicating that a 3 has been presented to the input. Just as for letters, this can be done by saying that a class exists, called class C_3 for example, in which all of the different types of 3 that were shown in Figure 1.1 belong. When a 3 is presented to the

Figure 1.1 Examples of the number 3

input, the machine responds by indicating that a member of class C_3 has been presented.

This example can be taken a bit further. Figure 1.2 shows a block diagram of the machine. The input will be a digitised version of an image, divided into 8 × 8 pixels, each with a value of 0 or 1 representing black or white respectively. This produces quite a crude representation of the number 3 but should be enough to be able to distinguish the 10 different numerals. The output is a lamp which indicates which number has been identified as the input. So we have a 64-input 10-output device.

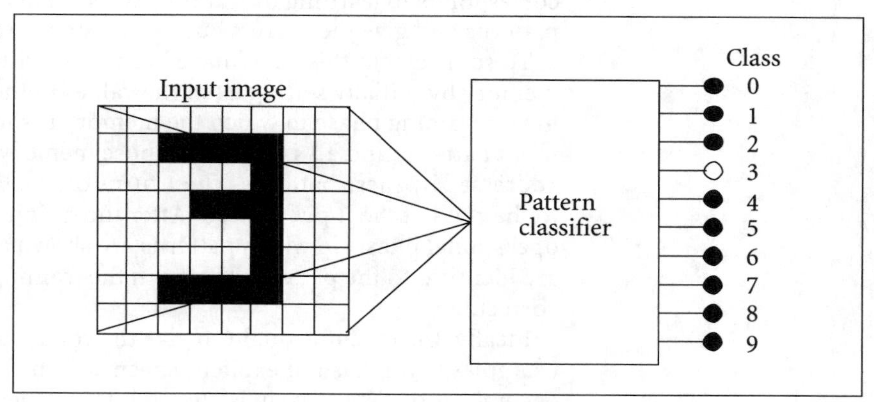

Figure 1.2 Block diagram of a pattern classifier

The aim is to design the device so that if a number appears at the input the corresponding lamp will light at the output. If the image at the input is of anything other than a number, no lights are lit. This represents a pattern classifier which can classify the 10 different numerals.

Probably the simplest conceptual solution to this problem would be to use an electronic memory. As there are 64 inputs, there are a possible 2^{64} different input patterns or 1.8×10^{19}, so a memory with 1.8×10^{19} locations would be needed. At each location there must be 10 bits, each bit representing one of the 10 different output values. Even if 1 Gbit memory chips were available, about 170 billion of them would still be needed!

To program the memory, every possible input combination would have to be looked at, a decision made as to whether it is a number, and, if it is a number, what number? Assuming that one pattern could be processed every second, then without interruption, it would take about 600 000 000 000 years to program it!

So, although this solution is conceptually simple, it is clearly beyond any kind of practical implementation. It does, however, represent a pattern classifier which is not a neural network.

1.3 Learning and generalisation

Neural networks are pattern classifiers, but not all pattern classifiers are neural networks. The next step is therefore to introduce some other features that distinguish neural networks from other types of classifiers. Probably the two most significant properties of neural networks are their ability to learn and to generalise.

First of all, learning. The solution to the numeral classifier problem using the electronic memory showed that it would take far too long to program the memory by going through every possible input combination. It would be better if we could program the machine by showing it a smaller number of examples while indicating the correct response. This corresponds to learning or training: the system learns to recognise certain patterns and give the correct output response to them.

To some extent this could have been done with the original electronic memory by initially setting all of the values in the memory to 0. Then, have a training phase in which the memory is shown examples of the input patterns and a 1 is placed in those memory locations that are addressed by these patterns. The 1 turns on the lamp which corresponds to the class of the input pattern. After the training phase comes the operational phase, in which patterns are shown again. If these patterns are identical to the patterns shown in the training phase the output will be correct.

Ideally, the machine ought to give the correct response even to examples that it has not explicitly been shown. This property is called generalisation. The system is able to infer the general properties of the different classes of patterns from the examples given. Neural networks are capable of doing this to some extent – if they are operating correctly they will respond to patterns that are very similar to the patterns that they have been shown in the training phase. In the case of the numeral classifier this means that the neural network can be shown examples of numbers, and it should correctly classify similar numbers which are not necessarily identical to the original examples.

In order to explain how this property of generalisation comes about, the detailed structure of the networks will have to be described. This is introduced in the next section.

To summarise, a neural network should be able to:

- classify patterns
- be small enough to be physically realizable
- be programmed by training, so it must have the ability to learn
- be able to generalise from the examples shown during training.

1.4 The structure of neural networks

In this section, some of the ways that have been developed of incorporating learning and generalisation into electronic devices will be described. The first is an example of a Boolean neural network which uses memory devices just as in the example described earlier, but this time in a way that allows generalisation. The second example is of the more biologically inspired neural network.

1.4.1 Boolean neural networks

An example of a system which classifies patterns, learns by example and which can generalise is the WISARD (Aleksander *et al.*, 1982). The term WISARD stands for **WI**lkie, **S**tonham and **A**leksander's **R**ecognition **D**evice. Contrary to the earlier statement that most neural networks are biologically inspired, the WISARD is not. In order to make this explicit, the WISARD and other devices like it that are built from conventional Boolean logic gates are often described as Boolean neural networks. In Chapter 4, the WISARD will be described in more detail, so at this point only a brief outline of its properties will be given.

Returning to the 8 × 8 image from the numeral classifier described earlier, first divide the image into smaller sub-images. For example, the image could be divided into four sections, as shown in Figure 1.3. Each section is connected to an electronic memory device – a random access memory or RAM. Each RAM is therefore connected to a quarter of all of the pixels, and so has 64/4 = 16 inputs. This means that each RAM will have 2^{16} or 64K locations, each of which can be addressed by the inputs. Also, instead of storing 10 bits per location, they only store 1 bit but have 10 different RAMs. The system then consists of 10 sets of four 64 kbit RAMs. Each set, called a discriminator, corresponds to one of the classes of pattern to be recognised.

The saving in memory size compared with the earlier memory-based solution is huge. Before, 2^{64} locations with 10 bits in each were required. Now, only 10 sets of 4×2^{16} locations are needed – nearly 70 million million times smaller!

The outputs of the 4 RAMs are connected to a summing junction. If all four RAMs produce a 1 output, the output of the summing junction is the maximum value of 4. If only three of the RAMs produce a 1 output, the other producing 0, then the output of the summing junction is 3, and the input pattern can be said to be 75% like the particular class.

Training consists of showing the system one of the patterns to be classified, an example of the number 3 say, and placing a 1 in the four memory locations in the 3-discriminator addressed by that input, as shown in Figure 1.3. The other memory devices in the other discriminators are unchanged.

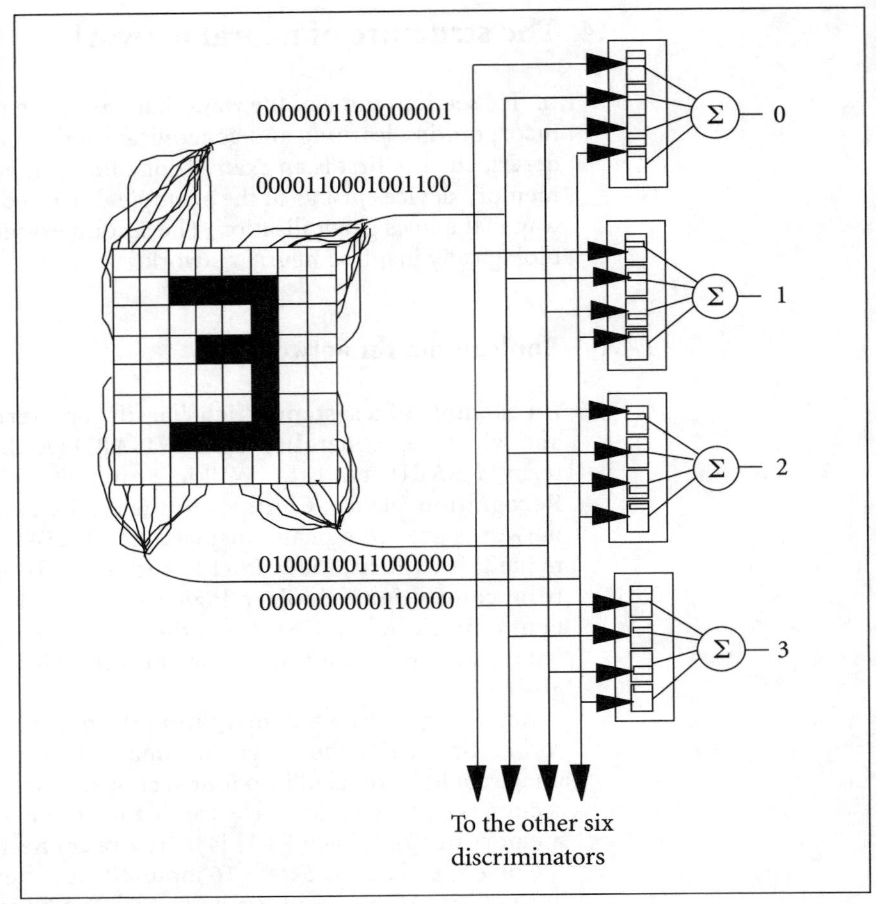

Figure 1.3 Training the WISARD to classify 3

If the system is shown several examples of the number 3, each example places 1s in the memory at different locations. Now if, after training, an input is shown which is identical to one of the training patterns, the output of one of the discriminators will be maximum. Some of the other discriminators may also produce an output because of common features between the patterns, but this should be less than the maximum.

An interesting feature of this system is that it can generalise. Figure 1.4(a) shows two examples of the number 3 which could be used to train this system. Although the system has learnt only two examples, there are 16 possible input patterns that would give a maximum response of 4, 14 of which have never been seen before. Four of these are shown in Figure 1.4(b). Although each of these is a very crude-looking 3, they are certainly more like a 3 than any other numeral. It is therefore appropriate that they should be classified as 3.

So the WISARD is an example of a pattern classifier that can be trained and which can generalise. It does have limitations, however, which will be discussed in Chapter 4.

Figure 1.4 (a) Two examples of the numeral 3; (b) patterns that would be recognised as a 3

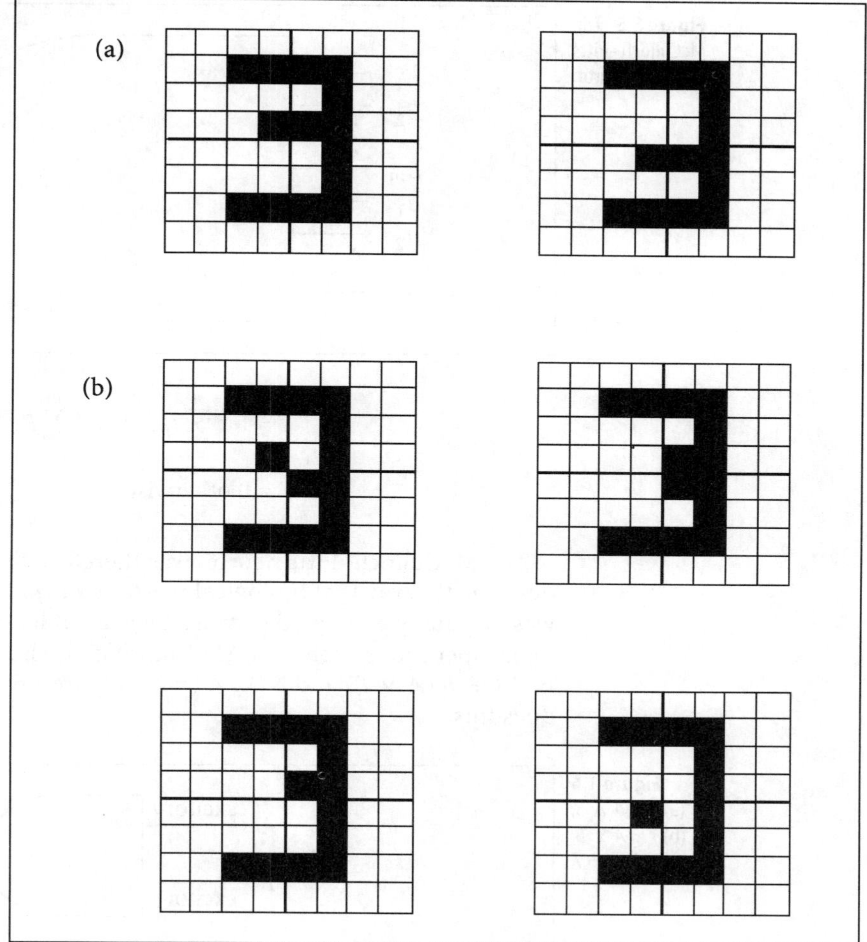

1.4.2 Biologically inspired neural networks

Most of the different types of neural network, apart from the Boolean ones, are composed of elements which are direct descendants of the model of a biological neuron created by McCulloch and Pitts (1943).

Figure 1.5 shows a diagram of a McCulloch–Pitts neuron. It has excitory inputs, *E*, and inhibitory inputs, *I*. In simple terms, excitory inputs cause the neuron to become active or 'fire', and inhibitory inputs prevent the neuron from becoming active. More precisely, if any of the inhibitory inputs are active (often described in binary terms as 1), the output, labelled *Y*, will be inactive or 0. Alternatively, if all of the inhibitory inputs are 0, and if the sum of the excitory inputs is greater than the threshold, *T*, then the output is active or 1. Mathematically, this is expressed as:

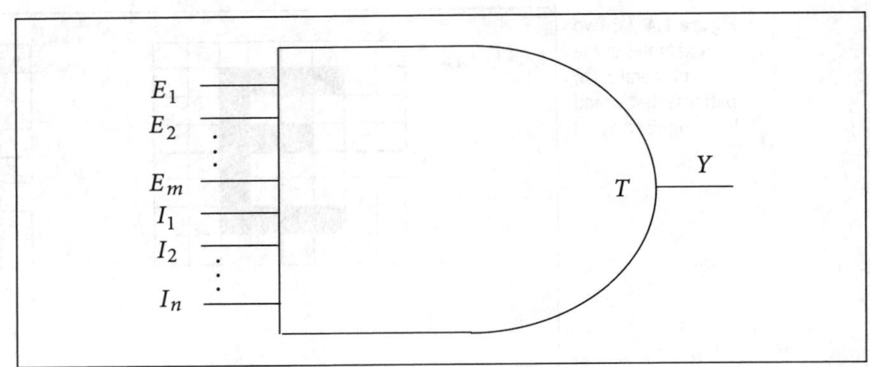

Figure 1.5 The McCulloch–Pitts neuron

$$Y = 1 \text{ if } \sum_{i=1}^{n} I_i = 0 \text{ and } \sum_{j=1}^{m} E_j \geq T$$

$$Y = 0 \text{ otherwise}$$

The McCulloch–Pitts neuron was therefore the first model that tried to describe the way that biological neurons work. Interest in these neurons was originally generated because they are able to perform basic Boolean logic operations such as A AND B, which is shown more formally as $A \wedge B$, A OR B ($A \vee B$) and NOT A ($\neg A$). Figure 1.6 shows how the neuron does this.

Figure 1.6 (a) $Y = A \wedge B$; (b) $Y = A \vee B$; (c) $Y = \neg A$

(a) A excitory, B excitory, threshold 2, output Y

(b) A excitory, B excitory, threshold 1, output Y

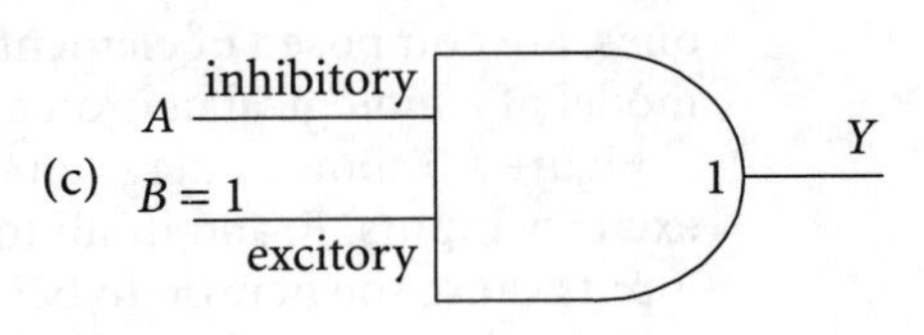

The basic building blocks of conventional computers are logic gates, which are electronic devices that perform the logical operation just described. So neurons can act as logic gates, and it was thought that the brain, which is made up of millions of neurons, must be a large computer. Therefore, if enough of these McCulloch–Pitts neurons were systematically organised, they would be able to mimic the functions of the brain. However, it was not realised immediately that these neurons lacked the ability to learn or to generalise.

Modifications have since been made to the original model. One of the first was made by von Neumann (von Neumann, 1956) who introduced the idea of making the inhibitory inputs negative. This meant that in order for the output of the neuron to be active, or fire, the sum of the excitory inputs minus the inhibitory inputs had to be greater than the value of the threshold. The next step was the introduction of 'weights' on each of the inputs. These weights are real numbers which multiply the value at the input, and so weight the inputs before they are summed.

One of the earliest neural networks that incorporated weighted inputs was the ADALINE developed by Widrow and Hoff (Widrow and Hoff, 1960). The name ADALINE is derived from the words **ADA**ptive **LI**near **NE**uron, or later **ADA**ptive **LIN**ear **E**lements. Figure 1.7 shows one of these elements.

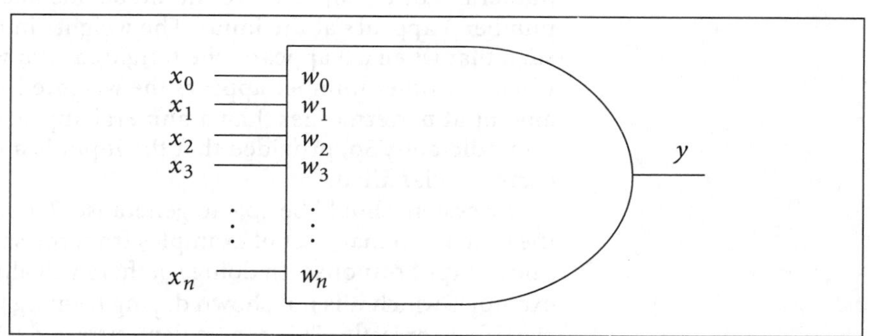

Figure 1.7 An element of ADALINE

Each element, or neuron, has several inputs which can take the value of +1 or –1. This differs from the McCulloch–Pitts neuron, which had input values of 1 or 0. Each input, x, has a weight, w, associated with it which gives some indication of the importance of that input. The weights can be positive or negative and have real values. When a pattern is presented at the inputs, the weighted sum, called the net input, *net*, is found by multiplying the inputs with their corresponding weights and adding all of the products together.

$$net = \sum_{i=0}^{n} w_i x_i$$

The additional input and weight, x_0 and w_0, provide a constant offset with x_0 set permanently to +1. This replaces the threshold that was used in the McCulloch–Pitts neuron.

The value of *net* is transformed into the value at the output, y, via a non-linear output function. This function gives an output of +1 if the weighted sum is greater than to 0 and –1 if the sum is less than or equal to 0. This sort of non-linearity is called a hard-limiter, which is defined as:

$$y = +1 \text{ if } net > 0$$

$$y = -1 \text{ if } net \leq 0$$

The main problem associated with this device and other neural networks is in the selection of the values for the weights. The details of the selection process will be discussed in depth in Chapter 2. Just accept that the weights are chosen such that the output is +1 or –1 as appropriate to the corresponding input pattern.

Training the neural network involves showing it a set of examples of the input patterns that you want to be classified, and allowing the weights to settle in such a way that all of the examples in the training set will produce the correct response. A single neuron will be able to produce an output of +1 or –1 which corresponds to the input pattern belonging to a particular class or not.

If the 0 to 9 numeral classification problem is to be implemented using an ADALINE, then at least 10 neurons are required, one per numeral. For example, there should be one neuron that fires when a number 3 appears at the input. The weights in that neuron will be set such that when a 3 appears, the weighted sum will be greater than 0 and when any other number appears the weighted sum will be less than 0. If any input pattern other than a numeral appears, the system will behave unpredictably. So, provided that the input is a number, it should be correctly classified.

The system should be able to generalise. This is achieved by adjusting the weights so that a set of examples from the same class produce the same output response. In doing so, there will almost certainly be other examples which it is not shown during training which will also get classified correctly. This can be demonstrated with a simple example, the number 3 again.

Figure 1.8 shows an example of a 3. The numbers +1 and –1 have been used to represent black and white respectively.

Figure 1.8 The number 3

–1	–1	–1	–1	–1	–1	–1	–1
–1	–1	+1	+1	+1	+1	–1	–1
–1	–1	–1	–1	–1	+1	–1	–1
–1	–1	–1	+1	+1	+1	–1	–1
–1	–1	–1	–1	–1	+1	–1	–1
–1	–1	–1	–1	–1	+1	–1	–1
–1	–1	+1	+1	+1	+1	–1	–1
–1	–1	–1	–1	–1	–1	–1	–1

One way that a neural network could be trained to recognise this 3 is to assign a value of –1 to the weights that correspond to inputs which are white, and +1 to the weights corresponding to inputs that are black. Since white is represented by –1 and black by +1, each of the weighted inputs are either $-1 \times -1 = 1$ or $+1 \times +1 = 1$, and the value of *net*, the sum of the weighted inputs, is 64 when presented with a perfect 3. If the offset, w_0, is set to –63, the weighted sum is 64 – 63 = 1, which, being greater than 0, gives an output of +1 after passing through the hard-limiter. Now, if any other input pattern appears, the weighted sum will be at most –1, and could be as low as –127, so the output is –1 for all other patterns. The original 3 produces a +1 output and anything else produces –1.

What if a slightly different 3 is to be recognised, such as an image with one bit corrupted (black instead of white, or vice versa)? Figure 1.9 shows an example of this.

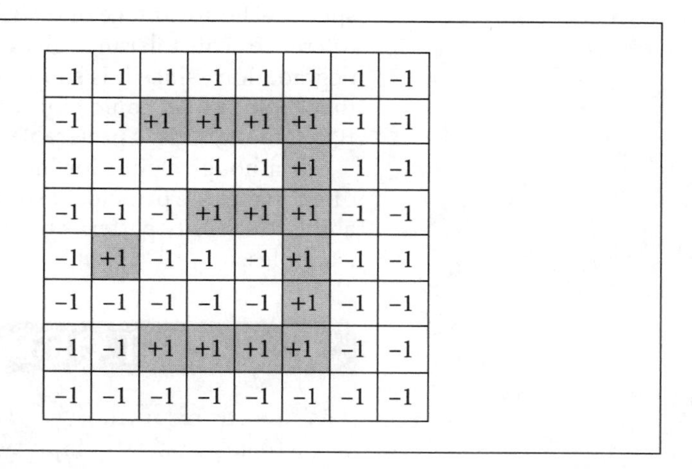

Figure 1.9 A 3 with one bit corrupted

The original 3 with one bit corrupted would produce a weighted sum of 62. This is found by taking the sum of all the instances of the weights and the corresponding inputs having the same value, either +1 or –1, and the one instance where the weight and the corresponding input are different giving a weighted sum of $(63 \times +1) + (1 \times -1)$, so that the total value for the sum is 62. When the offset, $w_0 = -63$, is added the total is –1, which is less than 0 and so produces a –1 output.

If the value of the offset, w_0, is changed to –61, the value of *net* for the corrupted 3 becomes 1, and the output is +1. The original 3 input would produce a *net* value of 3 with this new value of w_0, which still gives the correct output of +1 after the *net* value has been through the hard-limiter. So, by changing the value of w_0, the original image of 3 and ones with 1 bit error are correctly classified. Clearly, the system has not actually been shown all of the possible images with 1 bit error, but it has been able to generalise from the one example.

CHAPTER SUMMARY

This chapter started with the statement that neural networks produce output patterns when presented with input patterns, and then described one of its major applications as a pattern classifier. It showed that they also have the ability to learn and to generalise. This means that they can be taught to classify patterns by showing them good representative examples.

There are other properties of neural networks which have not been mentioned but which could be very important. The first is that they tend to work in parallel, with the work being distributed among the processing elements. This has led to another term that is used to describe neural networks, namely parallel distributed processors.

The advantage of being parallel is the potential for very high processing speeds. The advantage of distributed processing is a certain degree of tolerance. This tolerance gives the network the property called 'graceful degradation', since it is supposed that if part of the network malfunctions, the whole system could still continue to operate, albeit less than perfectly. It is said that this is preferable to complete failure.

So, although claims for increased speed and graceful degradation are often made, the principal feature of a neural network still remains its ability to classify patterns based on information that it has learnt from examples.

SELF-TEST QUESTIONS

1. Electronic memory is to be used as a pattern classifier, with binary images as inputs and lamps which can be on or off to represent the pattern classes as output. If the images to be recognised consist of N pixels, and there are M classes of patterns, how much electronic memory is needed and how many years would it take to program the memory if each location is set at a rate of 1 per second?
2. Let's assume that a WISARD is trained by showing it the two patterns that are given in Figure 1.4(a). Although connections would normally be random, let's just assume that the image is divided evenly into four quadrants for the moment. How many 8×8 patterns would be recognised as a 3 with the maximum score of 4?
3. An ADALINE is set up with the weights shown in Figure 1.8. How low would the offset have to be made in order for the ADALINE to still recognise a 3, even when there are P pixels that are corrupted?

SELF-TEST ANSWERS

1 Each pixel is an input, so there are N address inputs to the memory. This means that the memory has to have 2^N locations that it can address. If there are M classes, and each is represented by a single bit, then each memory location needs to hold M bits. So the total number of bits is:

$$\text{Number of bits} = M \times 2^N$$

The number of seconds in a year is $60 \times 60 \times 24 \times 365.24 = 31\,556\,736$ seconds. If the memory is programmed at 1 pattern per second, and the total number of patterns is 2^N, then the time taken in years is:

$$\text{Time needed} = 2^N/31\,556\,736 \text{ years}$$

2 For convenience, the two images will be referred to as image A (the one on the left) and image B (the one on the right). Similarly, the quadrants will be referred to as quadrants 1 to 4, starting at the top left and going around clockwise. So the bottom right quadrant of the image on the right can be referred to as B3.

The two images differ around the central four pixels. If we look at the top-left quadrant (1) in both images, then an image would be accepted if the bottom-right pixel is black or if it is white. Similarly, in all of the quadrants there is one pixel which is black in image A

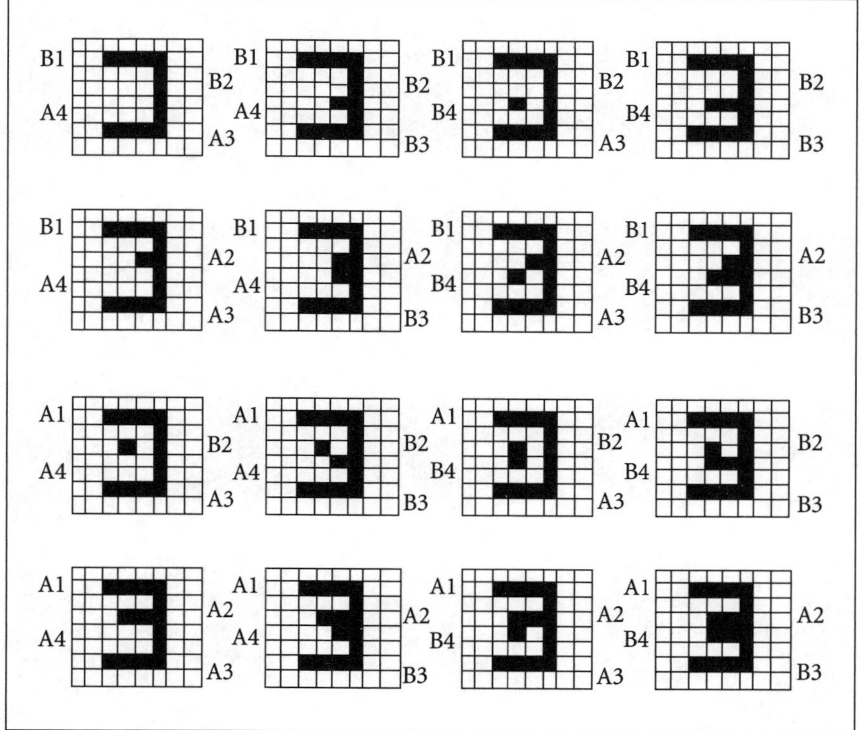

Figure 1.10 The 16 possible input patterns that give a maximum output response

and white in image B or vice versa. Because there are two versions of each quadrant, and there are four quadrants, there are a total of 24 combinations of quadrants. This means that there are a total of 16 possible patterns, as shown in Figure 1.10. Also shown in the figure is the name of the matching quadrant.

3 When the offset is −63, the weighted sum, *net*, for a perfect 3 is 1. If one pixel is corrupted, the value of *net* falls to −1, that is it drops by 2. If two pixels are corrupted it drops by another 2 to −3. So if P pixels are corrupted the weighted sum, *net*, drops by $2P$. If we still want to recognise these corrupted images, we have to drop the value of the offset by an equivalent amount. So the value of the offset becomes:

$$\text{offset} = -63 + 2P$$

CHAPTER 2

ADALINE

CHAPTER OVERVIEW

This chapter describes the ADALINE in more detail, particularly how it learns. It explains:

- how weight adjustment is achieved in an ADALINE
- what is meant by Hebbian learning
- what is meant by the delta rule

2.1 Training and weight adjustment

When a neural network, such as the ADALINE, is in its training or learning phase, there are three factors to be taken into account:

- the inputs that are applied are chosen from a training set where the desired response of the system to these inputs is known
- the actual output produced when an input pattern is applied is compared with the desired output and used to calculate an error
- the weights are adjusted to reduce the error

This kind of training is called supervised learning because the outputs are known and the network is being forced into producing the correct outputs. The alternative to this is unsupervised learning, where the outputs are not known in advance and the network is allowed to settle into suitable states. This will be covered in a later chapter.

The requirements of a supervised learning strategy are therefore a suitable weight adjustment mechanism and a suitable error function. Most attempts to find a mechanism for weight adjustment have been based on a statement that was made by Donald Hebb (Hebb, 1949), and consequently are known as Hebbian learning methods. Hebb's original statement was:

> When an axon of cell A is near enough to excite a cell B and repeatedly or persistently takes place in firing it, some growth process or metabolic change takes place in one or both cells such that A's efficiency, as one of the cells firing B, is increased.

The interpretation of this statement can be summarised by the following rule, which is often called Hebb's rule – increase the value of a weight if the output is active when the input associated with that weight is also active.

Numerically, this statement can be interpreted in several different ways. One way is to assume that the inputs, x, and outputs, y, are binary valued (0,1) and a weight, w, is incremented only when the output is 1 and the corresponding input is 1. If the weights are initially set to 0, their value after a set of training patterns have been applied to the inputs equals:

$$w_i = \sum_{p=1}^{P} x_{ip} y_p$$

where P is the number of patterns in the training set. This can be rewritten in matrix form as the product between the matrix of the input patterns and the matrix of the corresponding output values.

$$[W] = [X]^t[Y]$$

where $[X]^t$ is the transpose of the matrix [X].

For example, an ADALINE is given the four different input and output combinations of the 2-input AND function, $y = x_1 \wedge x_2$, as a training set. The truth table for this example, which shows all of the possible input combinations and the corresponding output values, is:

p	x_1	x_2	y
1	0	0	0
2	0	1	0
3	1	0	0
4	1	1	1

If the network is trained by presenting it with input and output pairs in the same order as the truth table, starting from the top, then the first input pattern that is used ($p = 1$) is 0 0, and the corresponding output pattern is 0. The arrays [X] and [Y] are therefore:

$$[X]=\begin{bmatrix}0 & 0\\0 & 1\\1 & 0\\1 & 1\end{bmatrix} \quad [X]^t=\begin{bmatrix}0 & 0 & 1 & 1\\0 & 1 & 0 & 1\end{bmatrix} \text{ and } [Y]=\begin{bmatrix}0\\0\\0\\1\end{bmatrix}$$

$$[W]=\begin{bmatrix}0 & 0 & 1 & 1\\0 & 1 & 0 & 1\end{bmatrix}\begin{bmatrix}0\\0\\0\\1\end{bmatrix}=\begin{bmatrix}1\\1\end{bmatrix}$$

So, $w_1 = 1$ and $w_2 = 1$. Now, if the network is shown the same training set, the corresponding weighted sum, *net*, would be:

x_1	x_2	*net*	y
0	0	0	0
0	1	1	0
1	0	1	0
1	1	2	1

Assuming that the output is a hard-limiter as described in Chapter 1, and that an offset of –1.5 is applied, the correct values for the output response are obtained, as shown in the previous table.

Examples of this sort, two-input logic functions, will be used to illustrate the various learning algorithms throughout this chapter. The reason for this is that it should be possible to calculate the weights by hand for such small examples. However, it should be stressed that these simple problems are not representative of the sort of problems which would normally be tackled using a neural network, mainly because the training set is completely specified. The network is shown all possible input patterns and only has to be able to give the correct response to these inputs. The network does not have to generalise in order to correctly classify new unseen inputs.

The use of the binary values 0 and 1 does not always produce a good result using this weight updating mechanism, however, as can be shown with the function $y = x_1 \wedge \neg x_2$. The truth table for this function is:

p	x_1	x_2	y
1	0	0	0
2	0	1	0
3	1	0	1
4	1	1	0

The weights are calculated using the product of the matrices as before.

$$[W] = \begin{bmatrix} 0 & 0 & 1 & 1 \\ 0 & 1 & 0 & 1 \end{bmatrix} \begin{bmatrix} 0 \\ 0 \\ 1 \\ 0 \end{bmatrix} = \begin{bmatrix} 1 \\ 0 \end{bmatrix}$$

When the network is shown the same training set again, the corresponding weighted sum and output are (assuming an offset of –0.5):

x_1	x_2	net	y	
0	0	0	0	
0	1	0	0	
1	0	1	1	
1	1	1	1	← error

The reason for this error is that the learning rule described so far only allows increases to the values of the weights. A more general approach would be also to allow the value of the weights to decrease. One way of achieving this is to decrease the value of a weight when the corresponding input is active and the output is not active, or vice versa. This can easily be achieved with the use of +1 and −1 instead of 1 and 0 again, as in the ADALINE, and then calculating the weights as the product between the matrix of input patterns and the corresponding matrix of output responses. For example, the truth table for the function $y = x_1 \wedge \neg x_2$ becomes:

p	x_1	x_2	y
1	−1	−1	−1
2	−1	+1	−1
3	+1	−1	+1
4	+1	+1	−1

Using matrices to find the weights using Hebbian learning gives:

$$[\mathrm{W}] = \begin{bmatrix} 1 & 1 & +1 & +1 \\ 1 & +1 & 1 & +1 \end{bmatrix} \begin{bmatrix} 1 \\ 1 \\ +1 \\ 1 \end{bmatrix} = \begin{bmatrix} +2 \\ 2 \end{bmatrix}$$

Applying the input patterns again with the weights set to the values just calculated gives:

x_1	x_2	net	y
−1	−1	0	0
−1	+1	−4	0
+1	−1	4	1
+1	+1	0	0

The correct output is obtained with a suitable offset and a hard-limiter.

This form of Hebbian learning does not always successfully train the network. It can only successfully train if the inputs are orthogonal or linearly independent (Rumelhart and McClelland, 1986). This is because

Hebbian learning takes no account of the actual value of the output, only the desired value. This limitation can be overcome if the weights are adjusted by an amount which depends upon the error between the desired and actual outputs. This error is called delta, δ, and the new learning rule is called the delta rule.

2.2 The delta rule

In the delta rule, the adjustments to the weights are made in such a way that the error between the desired output and the actual output is reduced. For example, if the output is +1 and the desired output is –1 the weights are adjusted so that the new value of the product of $w_i x_i$ is less than it was previously. This means that if x_i is negative the weight is increased, or if x_i is positive the value of the weight is decreased.

Intuitively, the adjustment to the value of w_i, Δw_i, should be proportional to x_i and to the error:

$$\Delta w_i \propto x_i \delta$$

where δ is the error between the desired output, d, and the actual output, y.

Widrow and Hoff developed a method of weight updating based on these assumptions. This is what has become known as the delta rule, but is also referred to as the Widrow–Hoff rule and the least mean square (LMS) rule. The latter name refers to the fact that Widrow and Hoff's method of weight updating minimises the squared error between the desired output and the actual output. In order to achieve this, a modification to the original intuitively derived equation has to be made. This is due to the output of the ADALINE passing through a non-linear function, namely the hard-limiter. In order to ensure that the changes to the weights result in a neuron that performs correctly, Widrow and Hoff had to modify the definition of δ so that it becomes:

$$\delta = d - net$$

In the case of the ADALINE, d and x have values of +1 or –1, and *net* is the value of the weighted sum. The squared error can be minimised using the following equation, which is derived in Appendix A.

$$\Delta w_i = \frac{\eta}{P} \sum_{p=1}^{P} \delta_p x_{ip}$$

Training an ADALINE using the delta rule therefore consists of finding the mean values of $x\delta$ for each of the inputs over the whole training set. The weights are then adjusted by this amount times η, which is a constant that is decided by the user.

Three additional points need to be included before the learning rule can be used:

- The constant, η, has to be decided. The original suggestion for the ADALINE was that η is made equal to $1/(n+1)$, where n is the number of inputs. The effect of adjusting the weights by this amount is to reduce the error for the current input pattern to zero. However, in practice if η is set to this value the weights rarely settle down to a constant value and a smaller value is generally used.
- The weights are initially set to a small random value. This is to ensure that the weights are all different.
- The offset, w_0, gets adjusted in the same way as the other weights, except that the corresponding input x_0 is assumed to be +1, as if the offset was a weighted input where the input is permanently stuck at a value of +1.

Now we can see what happens when this rule is applied to the problem described earlier of $y = x_1 \wedge \neg x_2$. The following solution uses a value of 0.1 for η. The weights are initially set to −0.12, 0.4 and 0.65.

First, the input pattern +1 −1 −1 is presented to the network and the weighted sum is −1.17, which goes through the hard-limiter to produce an output response of −1. The value of δ is found, where $\delta = d - net = 0.17$, and the values of $\eta\delta x_i$ calculated for each of the weights and stored. For convenience, these figures have been rounded to two places after the decimal point, so become $0.02x_i$.

The next input pattern, +1 −1 +1, is applied. This produces a weighted sum of 0.13, which, after passing through the hard-limiter, produces an output value of +1. The value of δ is found which is $d - net = -1.13$, from which $\eta\delta x_i$ is calculated and rounded to give $0.11x_i$. This continues until all of the four input patterns have been shown to the network.

x_0	x_1	x_2	w_0	w_1	w_2	net	d	$\eta\delta x_0$	$\eta\delta x_1$	$\eta\delta x_2$
+1	−1	−1	−0.12	0.40	0.65	−1.17	−1	0.02	−0.02	−0.02
+1	−1	+1	−0.12	0.40	0.65	0.13	−1	−0.11	0.11	−0.11
+1	+1	−1	−0.12	0.40	0.65	−0.37	+1	0.14	0.14	−0.14
+1	+1	+1	−0.12	0.40	0.65	0.93	−1	−0.19	−0.19	−0.19
							total	−0.14	0.04	−0.46
							mean	−0.04	0.01	−0.12

First pass through the training set

After the first pass through the training set the values of $\eta\delta x_i$ are added up to find the total for each weight, and the mean value found. This value is used to adjust the weights, so the weights become −0.16, 0.41 and 0.53 respectively. Next, the input patterns are presented to the network again, starting with the input pattern +1 −1 −1.

x_0	x_1	x_2	w_0	w_1	w_2	net	d	$\eta\delta x_0$	$\eta\delta x_1$	$\eta\delta x_2$
+1	-1	-1	-0.16	0.41	0.53	-1.10	-1	0.01	-0.01	-0.01
+1	-1	+1	-0.16	0.41	0.53	0.04	-1	-0.10	0.10	-0.10
+1	+1	-1	-0.16	0.41	0.53	-0.28	+1	0.13	0.13	-0.13
+1	+1	+1	-0.16	0.41	0.53	0.78	-1	-0.18	-0.18	-0.18
Second pass through the training set							total	-0.14	0.04	-0.44
							mean	-0.04	0.01	-0.11

Training continues by cycling through all of the test patterns until the output is correct for all inputs. When (and if) this happens, the weights stop changing, and the ADALINE is said to have converged to a solution. This happens when the weights have the values of –0.5, 0.5 and –0.5 respectively, so that the total, and hence the mean, change to the weights is 0.

x_0	x_1	x_2	w_0	w_1	w_2	net	d	$\eta\delta x_0$	$\eta\delta x_1$	$\eta\delta x_2$
+1	-1	-1	-0.50	0.50	-0.50	-0.50	-1	-0.05	0.05	0.05
+1	-1	+1	-0.50	0.50	-0.50	-1.50	-1	0.05	-0.05	0.05
+1	+1	-1	-0.50	0.50	-0.50	0.50	+1	0.05	0.05	-0.05
+1	+1	+1	-0.50	0.50	-0.50	-0.50	-1	-0.05	-0.05	-0.05
							total	0.00	0.00	0.00

The network has successfully found a set of weights that produces the correct outputs for all of the patterns.

Ideally, the mean value of $x\delta$ should be found over the whole training set when training. In practice, the weights are nearly always adjusted using the value of $x\delta$ at each showing of a training pattern, even though it has not been proven that this is correct (Minsky and Papert, 1989, p. 263; Rumelhart and McClelland, 1986, p. 324). This can be demonstrated using the same example and starting conditions as before.

When the input pattern +1 –1 –1 is presented to the network, the weighted sum is –1.17, which goes through the hard-limiter to produce an output response of –1. The weights are adjusted by an amount $\Delta w_i = \eta x_i \delta$, where $\delta = d - net = 0.17$, so $\Delta w_i = 0.017x_i$. This is rounded to two places after the decimal point, $\Delta w_i = 0.02x_i$. The weights are adjusted by this amount, and the next input pattern, +1 –1 +1, applied. This continues until all of the four input patterns have been shown to the network.

x_0	x_1	x_2	w_0	w_1	w_2	*net*	*d*	Δw_0	Δw_1	Δw_2
+1	−1	−1	−0.12	0.40	0.65	−1.17	−1	0.02	−0.02	−0.02
+1	−1	+1	−0.10	0.38	0.63	0.15	−1	−0.12	0.12	−0.12
+1	+1	−1	−0.22	0.50	0.51	−0.23	+1	0.12	0.12	−0.12
+1	+1	+1	−0.10	0.62	0.39	0.91	−1	−0.19	−0.19	−0.19

After the four input patterns are shown, the process is repeated starting from the first pattern. Continue cycling through the inputs until the output is correct for all inputs and the weights stop changing. In this example, the sequence starts to repeat when it gets to:

x_0	x_1	x_2	w_0	w_1	w_2	*net*	*d*	Δw_0	Δw_1	Δw_2
+1	−1	−1	−0.50	0.47	−0.56	−0.41	−1	−0.06	0.06	0.06
+1	−1	+1	−0.56	0.53	−0.50	−1.59	−1	0.06	−0.06	0.06
+1	+1	−1	−0.50	0.47	−0.44	0.41	+1	0.06	0.06	−0.06
+1	+1	+1	−0.44	0.53	−0.50	−0.41	−1	−0.06	−0.06	−0.06

The solution, therefore, never settles to constant values, but oscillates around the values of −0.5, 0.5 and −0.5.

The 'cost' of the delta rule is the number of cycles or iterations required to converge to a solution. This cannot be predicted in advance and it cannot be guaranteed to converge at all for some problems. When the weights are updated using the mean of the error terms, the theory predicts that the method will converge.

It is worth repeating that in this example all of the input combinations are known. In general, only a small subset of the total number of patterns would be shown during training, and although the weights are optimised for these patterns, it is also hoped that the error produced when new patterns are shown will be small.

One feature of the ADALINE system is the fact that the weights continue to adjust even when the correct solution has been found. The output from the non-linearity will be the correct value of +1 or −1, but the value of the weighted sum will probably not be exactly +1 or −1, so the weights will have small changes made to them.

It also means that if the ADALINE is shown a pattern that is quite different from the training set, a new class even, then the weights will be adjusted to accommodate this new class with the 'minimum disturbance', so that in principle the other patterns that it has already learnt should still be correctly classified (Widrow and Winter, 1988).

To summarise, the outputs of the ADALINE are found by taking the weighted sum of the inputs, and then passing the weighted sum through a non-linearity such as a hard-limiter. During the training or learning phase, the weights are adjusted so that the outputs correspond as closely as possible to the desired output values for the training set. A number of rules exist for adjusting the weights, all of which are variations on the principle described by Hebb. The rule that has been described here is called the Widrow–Hoff, LMS or delta rule.

All of the discussion so far has been concerned with single neurons, or more precisely a single layer of neurons. If the output is considered as a function of the inputs, then unfortunately it can be shown that not all functions can be implemented using only a single layer. This means that not all classifications are possible. Those that can are called linearly separable, and there will be more about this in the next chapter.

If an attempt is made to implement a non-linearly separable function on a single ADALINE, the solution that it arrives at will contain errors. However, because the delta rule minimises the mean squared error, the solution will be a kind of 'best fit', in the sense that it will be the best that can be achieved using just a single neuron.

One way of overcoming this problem is to have multiple layers of interconnected ADALINEs. However, the delta rule for training the ADALINE is insufficient for multi-layered networks, and research is still being done into developing new learning rules.

A solution that has been tried is to have a single layer of ADALINEs feeding into a logic device such as an AND gate. Designs using these methods are described by Widrow and Winter (Widrow and Winter, 1988). These multi-layered networks are called MADALINEs, the M standing for 'multi'.

2.3 Input and output values

In the examples that have been given, the binary values of +1 and −1 have been used for the inputs and outputs. However, these may not be as appropriate, for example, as in logic design problems which require 0 and 1. If this is the case, then it is quite straightforward to convert the networks as follows:

Outputs if a 0,1 output is required then simply replace the hard-limiter with one which gives a 0,1 output, as shown in Figure 2.1. Since the delta rule uses the weighted sum for comparison, the

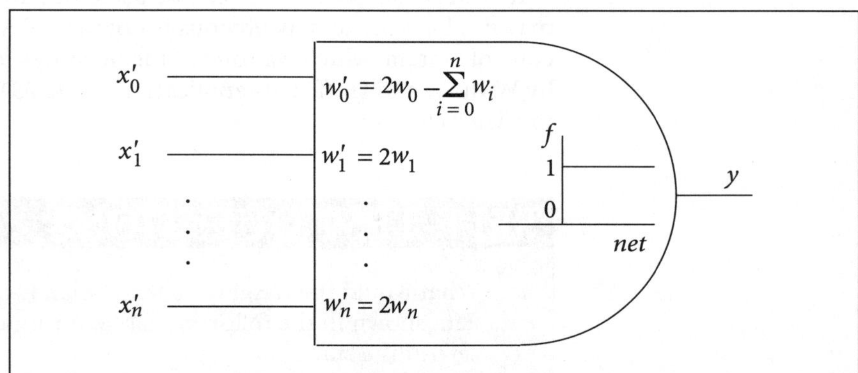

Figure 2.1
ADALINE with 0,1 inputs and outputs

output of the hard-limiter does not influence the values of the weights.

Inputs — if 0,1 inputs are required then they can be converted to −1,+1 inputs by multiplying by 2 then subtracting 1. The same can be achieved by altering the weights. If the value of the inputs x_i are −1,+1 and the value of the alternative inputs x'_j are 0,1 then the values can be interchanged using the equation $x_i = 2x'_j - 1$, and the weighted sum is therefore:

$$\sum_{i=0}^{n} w_i x_i = \sum_{i=0}^{n} w_i (2x'_i - 1)$$

$$\sum_{i=0}^{n} w_i x_i = \sum_{i=0}^{n} (2w_i x'_i - w_i)$$

$$\sum_{i=0}^{n} w_i x_i = \sum_{i=0}^{n} (2w_i) x'_i - \sum_{i=0}^{n} w_i$$

Thus, if the value of all the weights including w_0 are doubled and an amount equal to is subtracted from the new value of w_0, the neuron will behave with 0,1 inputs as it did with −1,+1 inputs, as shown in Figure 2.1.

Therefore, if data is specified using 0 and 1, the simplest thing to do is to convert it to −1 and +1 and train the network as normal. Alternatively, the data could be kept as 0 and 1 and the delta rule modified by replacing x_i by $(2x'_j - 1)$.

CHAPTER SUMMARY

In this chapter you have been shown the delta rule developed by Widrow and Hoff for the ADALINE. It adjusts the weights so that the mean squared error between the weighted sum and the desired output is minimised. It is limited, however, to single-layer networks, which can only implement a small proportion of functions. Some methods have been developed for multi-layered ADALINEs, or MADALINEs, but they are complex and time-consuming.

It is clear that ADALINEs can be used for pattern recognition, but what may not be immediately obvious is how an ADALINE can be used in a control system, which was one of the purposes for which it was developed by Widrow and Hoff. This application of the ADALINE will be discussed in a later chapter.

SELF-TEST QUESTIONS

1 What would the weights be if Hebbian learning is applied to the data shown in the following table? Assume that the weights are all zero at the start.

p	x_1	x_2	y
1	0	0	1
2	0	1	1
3	1	0	0
4	1	1	1

With the weights that you've just found, what output values are produced with a threshold of 1?

2 What would the weights be if the pattern being learned is the one shown in the following table? Assume that the weights are all zero at the start.

p	x_1	x_2	y
1	−1	−1	+1
2	−1	+1	+1
3	+1	−1	−1
4	+1	+1	+1

With the weights that you've just found, what output values are produced with a threshold of −1?

3 A 2-input ADALINE has the following set of weights:

$$w_0 = 0.3,\ w_1 = -2.0,\ w_2 = 1.5$$

When the input pattern is:

$$x_0 = 1,\ x_1 = 1,\ x_2 = -1$$

and the desired output is 1.

(a) What is the actual output?

(b) What is the value of δ?

(c) Assuming that the weights are updated after each pattern and the value of η is $1/(n+1)$, what are the new values for the weights?

(d) Using these new values of weights, what would the output be for the same input pattern?

4 With η set to 0.5, calculate the weights (to one decimal place) in the following example after one iteration through the set of training patterns (a) updating after all the patterns are presented and (b) updating after each pattern is presented.

x_0	x_1	x_2	w_0	w_1	w_2	*net*	*d*	$\eta\delta x_0$	$\eta\delta x_1$	$\eta\delta x_2$
+1	–1	–1	–0.2	0.1	0.3	–0.6	+1	+0.8	–0.8	–0.8
+1	–1	+1					+1			
+1	+1	–1					–1			
+1	+1	+1					+1			

SELF-TEST ANSWERS

1 Initially the weights are set to 0. Starting with the first row, where p = 1, the weights become:

$$p = 1 \quad w_1 = 0 + x_1 \times y = 0 + 0 \times 1 = 0$$
$$w_2 = 0 + x_2 \times y = 0 + 0 \times 1 = 0$$
$$p = 2 \quad w_1 = 0 + x_1 \times y = 0 + 0 \times 1 = 0$$
$$w_2 = 0 + x_2 \times y = 0 + 1 \times 1 = 1$$
$$p = 3 \quad w_1 = 0 + x_1 \times y = 0 + 1 \times 0 = 0$$
$$w_2 = 1 + x_2 \times y = 1 + 0 \times 0 = 1$$
$$p = 4 \quad w_1 = 0 + x_1 \times y = 0 + 1 \times 1 = 1$$
$$w_2 = 1 + x_2 \times y = 1 + 1 \times 1 = 2$$

So you end up with $w_1 = 1$, $w_2 = 2$.

When these weights are used with the same set of input values, the resulting weighted sum, called net, is as follows:

p	x_1	x_2	*net*	y
1	0	0	0	0
2	0	1	2	1
3	1	0	1	0
4	1	1	3	1

If we set an arbitrary threshold of 1, which means that when the weighted sum is greater than 1 the output is 1, otherwise it is 0, the values for y as shown above can be produced. Only three of these values are correct.

2 Initially the weights are set to 0. Starting with the first row, where p = 1, the weights become:

$$p = 1 \quad w_1 = 0 + x_1 \times y = 0 + (-1 \times +1) = -1$$
$$w_2 = 0 + x_2 \times y = 0 + (-1 \times +1) = -1$$

$$p = 2 \quad w_1 = -1 + x_1 \times y = -1 + (-1 \times +1) = -2$$

$$w_2 = -1 + x_2 \times y = -1 + (+1 \times +1) = 0$$

$$p = 3 \quad w_1 = -2 + x_1 \times y = -2 + (+1 \times -1) = -3$$

$$w_2 = 0 + x_2 \times y = 0 + (-1 \times -1) = 1$$

$$p = 4 \quad w_1 = -3 + x_1 \times y = -3 + (+1 \times +1) = -2$$

$$w_2 = 1 + x_2 \times y = 1 + (+1 \times +1) = 2$$

So you end up with $w_1 = -2$, $w_2 = 2$.

With these values and a threshold of –1, the function looks like:

p	x_1	x_2	*net*	y
1	–1	–1	0	+1
2	–1	+1	+4	+1
3	+1	–1	–4	–1
4	+1	+1	0	+1

3 An ADALINE has the following set of weights:

$$w_0 = 0.3, w_1 = -2.0, w_2 = 1.5.$$

When the input pattern is:

$$x_0 = 1, x_1 = 1, x_2 = -1$$

the desired output is 1.

(a) The weighted sum, *net*, is:

$$net = \sum_{i=0}^{3} w_i x_i = x_0 \times w_0 + x_1 \times w_1 + x_2 \times w_2$$

$$net = (0.3 \times 1) + (-2.0 \times 1) + (1.5 \times -1) = -3.2$$

With this value for *net*, the actual output would be 0 since $net \leq 0$.

(b) $\delta = d - net$ where d is the desired output.

$$\delta = 1 - (-3.2) = 4.2$$

(c) The weights are updated by Δw_i which equals:

$$\Delta w_i = \eta \delta x$$

The value of η is $1/(n+1)$ where $n = 2$, so $\eta = 0.33$.

$$\Delta w_0 = 0.33 \times 4.2 \times 1 = 1.4$$

$$\Delta w_1 = 0.33 \times 4.2 \times 1 = 1.4$$

$$\Delta w_2 = 0.33 \times 4.2 \times -1 = -1.4$$

So:

$$w_0 = 0.3 + 1.4 = 1.7$$
$$w_1 = -2.0 + 1.4 = -0.6$$
$$w_2 = 1.5 - 1.4 = 0.1$$

(d) With these weights *net* is:

$$net = (1.7 \times 1) + (-0.6 \times 1) + 0.1 \times -1 = 1.0$$

With this value for *net*, the actual output would be 1 since $net > 0$. Note that the error, δ, would now be zero.

4 (a) Updating after all patterns are presented.

x_0	x_1	x_2	w_0	w_1	w_2	*net*	*d*	$\eta\delta x_0$	$\eta\delta x_1$	$\eta\delta x_2$
+1	−1	−1	−0.2	0.1	0.3	−0.6	+1	+0.8	−0.8	−0.8
+1	−1	+1	−0.2	0.1	0.3	0.0	+1	+0.5	−0.5	+0.5
+1	+1	−1	−0.2	0.1	0.3	−0.4	−1	−0.3	−0.3	+0.3
+1	+1	+1	−0.2	0.1	0.3	+0.2	+1	+0.4	+0.4	+0.4
							Total	+1.4	−1.2	+0.4
							Mean	+0.4	−0.3	+0.1

Therefore the weights get adjusted to $w_0 = +0.2$, $w_1 = -0.2$, $w_2 = +0.4$. After two more iterations, the weights converge to the solution where $w_0 = +0.5$, $w_1 = -0.5$ and $w_2 = +0.5$.

(b) Updating after each pattern.

x_0	x_1	x_2	w_0	w_1	w_2	*net*	*d*	$\eta\delta x_0$	$\eta\delta x_1$	$\eta\delta x_2$
+1	−1	−1	−0.2	0.1	0.3	−0.6	+1	+0.8	−0.8	−0.8
+1	−1	+1	+0.6	−0.7	−0.5	+0.8	+1	+0.1	−0.1	+0.1
+1	+1	−1	+0.7	−0.8	−0.4	+0.3	−1	−0.7	−0.7	+0.7
+1	+1	+1	+0.0	−1.5	+0.3	−1.2	+1	+1.1	+1.1	+1.1

Therefore the weights get adjusted to $w_0 = +1.1$, $w_1 = -0.4$, $w_2 = +1.4$.

Either way, the weights get adjusted to reduce the error between the desired output and the weighted sum. Although the weights have different values, in both cases the set of weights have the same signs and, if tried with the input data, would both give the correct response. In this second case, however, after a few iterations it is apparent that the values of the weights are oscillating around the values found in the previous case. Reducing the learning coefficient helps to improve this, but the values still oscillate. This is due, in part, to the rounding of the values to one decimal place, but is quite typical of this method.

CHAPTER 3

Perceptrons

CHAPTER OVERVIEW

This chapter introduces the perceptron, which can be connected in multiple layers and can consequently learn to classify any patterns. It explains:

- how a single neuron separates patterns using hyperplanes
- how a multi-layered perceptron can classify non-linearly separable patterns
- how a multi-layered perceptron can learn using back-propagation
- how to calculate the change to a weight in the network using back-propagation
- how to choose the architecture of a multi-layered perceptron

3.1 Single layer perceptrons

Perceptrons are the most widely used and best understood of all the different neural networks. The term was first used by Frank Rosenblatt (Rosenblatt, 1958) to describe a number of different types of neural network. It was his idea to take the functional description of how a neuron works and to implement it as an algorithm in software rather than trying to build a physical model of a neuron like the ADALINE.

The perceptron which became known as the Mark 1 perceptron is shown in Figure 3.1. It consists of three layers of units called the S (sensor), A (association) and R (response) units. Only the last of these, the R units, have adjustable weights, and consequently these units are the ones that have been paid the most attention. In fact, the term 'perceptron' has become synonymous with the units in the R layer. This is why this network would be considered to be a single-layered perceptron, as there is only one layer of elements with adjustable weights.

Briefly, the S layer consists of a digitized image, where 1 is used to represent white pixels and 0 for dark pixels. Groups of pixels are randomly connected to units in the A layer. These produce binary 0/1 outputs as a function of the inputs – in other words they are logical feature detectors. For example, one of the units might fire (produce a 1 output) when the majority of pixels connected to it are 1.

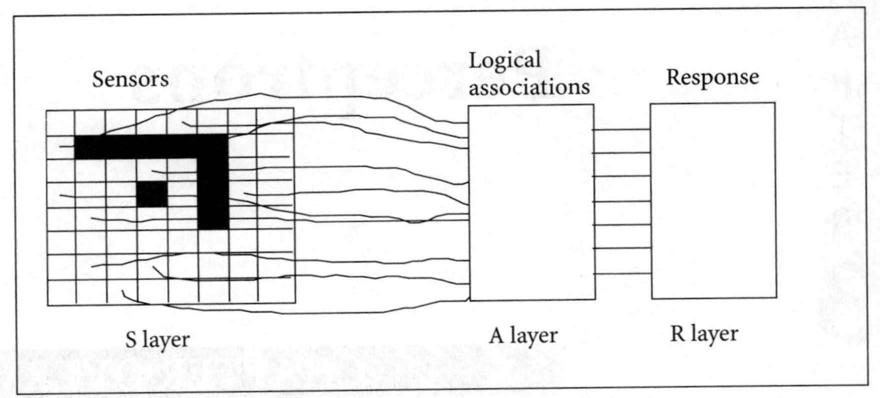

Figure 3.1 The Mark 1 perceptron

Finally, the outputs from the A layer are connected to units in the R layer, which behave very much like the ADALINE that was described in the previous chapter. They have a number of inputs, weights associated with those inputs, a weighted sum, *net*, is found, and finally an output is generated by a non-linear output function, such as a hard-limiter.

The operation of the response layer of the perceptron, once it has been trained, is much the same as the ADALINE. Inputs are weighted and summed, and then an output of either 0 or 1 is generated, depending on the value of the weighted sum.

In detail, an R unit of the perceptron and an ADALINE differ in a number of respects. The most obvious is that the inputs and outputs are binary values of 0 and 1 rather than −1 and +1.

Training is slightly different from the ADALINE, although it is still referred to as the delta rule. This time the error is measured as the difference between the desired output and the actual output from the non-linear output function. Another difference is that the weight adjustment is incremental, in the sense that only unit changes are made. So:

$$\Delta w_i = x_i \delta$$

The variable $\delta = d - y$, and since the values of d and y are 0 or 1, the value of Δw is either −1, 0 or +1.

The following example shows what happens for the function $y = x_1 \wedge \neg x_2$. Initially, the weights must be set to any value except 0. For this example, they have each been given a value of 1, which means that the weights will be integer values only. This is not necessary, but it makes the arithmetic easier. The first pass through the training set looks like this:

x_0	x_1	x_2	w_0	w_1	w_2	*net*	y	d	Δw_0	Δw_1	Δw_2
1	0	0	1	1	1	1	1	0	−1	0	0
1	0	1	0	1	1	1	1	0	−1	0	−1
1	1	0	−1	1	0	0	0	1	1	1	0
1	1	1	0	2	0	2	1	0	−1	−1	−1

Note that the weights are updated after each input pattern is presented to the perceptron, although they could have been updated using the mean value of the error, as described for the ADALINE.

After a number of cycles the error is reduced to 0 and the weights converge to 0, 2 and –2. The response of the network to the set of input patterns is:

x_0	x_1	x_2	w_0	w_1	w_2	*net*	y	d	Δw_0	Δw_1	Δw_2
1	0	0	0	2	–2	0	0	0	0	0	0
1	0	1	0	2	–2	–2	0	0	0	0	0
1	1	0	0	2	–2	2	1	1	0	0	0
1	1	1	0	2	–2	0	0	0	0	0	0

Since Rosenblatt's original perceptron, there have been many variations. These include:

- lifting the restriction that inputs are 0 or 1, allowing any real number instead
- including a learning coefficient, η, as in the ADALINE

The following example shows what happens when the learning coefficient, η, is 0.1. The weights are initially set to random non-integer values.

x_0	x_1	x_2	w_0	w_1	w_2	*net*	y	d	Δw_0	Δw_1	Δw_2
1	0	0	–0.12	0.40	0.65	–0.12	0	0	0	0	0
1	0	1	–0.12	0.40	0.65	0.53	1	0	–0.1	0	–0.1
1	1	0	–0.22	0.40	0.55	0.18	1	1	0	0	0
1	1	1	–0.22	0.40	0.55	0.73	1	0	–0.1	–0.1	–0.1

The system finally converges to the weights –0.32, 0.4 and –0.15, and the network responds to the set of input patterns as follows:

x_0	x_1	x_2	w_0	w_1	w_2	*net*	y	d	Δw_0	Δw_1	Δw_2
1	0	0	–0.32	0.40	–0.15	–0.32	0	0	0	0	0
1	0	1	–0.32	0.40	–0.15	–0.47	0	0	0	0	0
1	1	0	–0.32	0.40	–0.15	0.08	1	1	0	0	0
1	1	1	–0.32	0.40	–0.15	–0.07	0	0	0	0	0

The noticeable feature of this learning rule is that the weights are adjusted by a fixed amount, $\pm\eta$. This is in contrast to the ADALINE, where the amount of change to the weights was variable. However, in both the perceptron and the ADALINE the changes to the weights are quite small, so that the number of iterations needed can often be quite large.

The importance of perceptrons in the development of neural networks is partly due to the perceptron convergence theorem . This states that if a solution can be implemented on a perceptron, the learning rule will find a solution in a finite number of steps. The proof of this theorem is given in great detail in the book *Perceptrons* (Minsky and Papert, 1989).

3.1.1 Limitations of the single-layered perceptron

Just as for the ADALINE, only linearly separable functions can be implemented on a single-layer network, so the perceptron convergence theorem only applies to single-layered perceptrons.

The limitation of only being able to implement linearly separable functions can be overcome by using multi-layered networks. Just as for the ADALINE, the learning rule that has just been described and indeed, the output function, are inadequate to train multi-layered networks. The requirement of the delta learning rule for a single neuron is that an error can be measured at the output. What constitutes the desired value of the output of a perceptron which is not directly connected to the outside world, that is, one that is only connected to other perceptrons, is the major issue.

3.2 Linear separability

The function $y = x_1 \wedge \neg x_2$ could be implemented using a single perceptron, so this must be an example of a linearly separable function. The term comes from the fact that if a function is drawn in a particular way, the classification between inputs that produce a 1 output and those that produce a 0 output lie on either side of a line.

The function is drawn in what is called ‘pattern space’, where each of the inputs represents a coordinate system for that space and the set of coordinates give the position of the feature in that space. Figure 3.2 shows the pattern space for two binary variables.

Figure 3.2 shows that the two input variables, labelled x_1 and x_2, are the axes of the diagram. The four points marked represent the positions in the diagram where x_1 and x_2 are equal to the four different combinations of 0

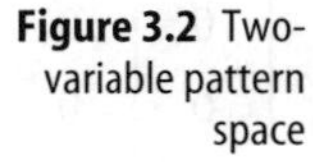
Figure 3.2 Two-variable pattern space

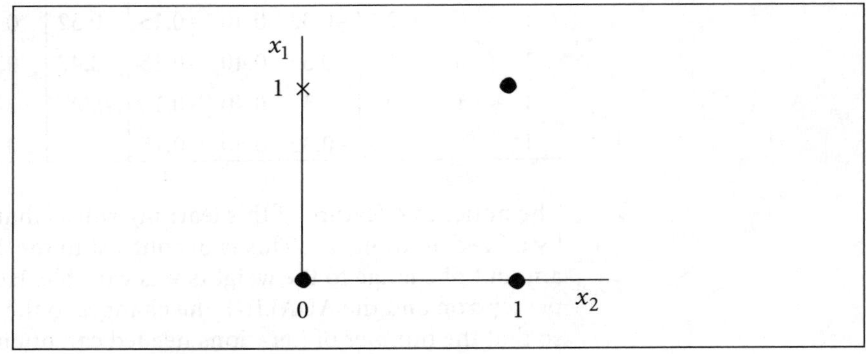

and 1. At each point the value of the output of the function $y = x_1 \wedge \neg x_2$ is marked as either a dot (for 0) or a cross (for 1).

It is possible to draw a line on this diagram which separates the 0s and 1s, as shown in Figure 3.3.

Figure 3.3 Separation of the 0s and 1s

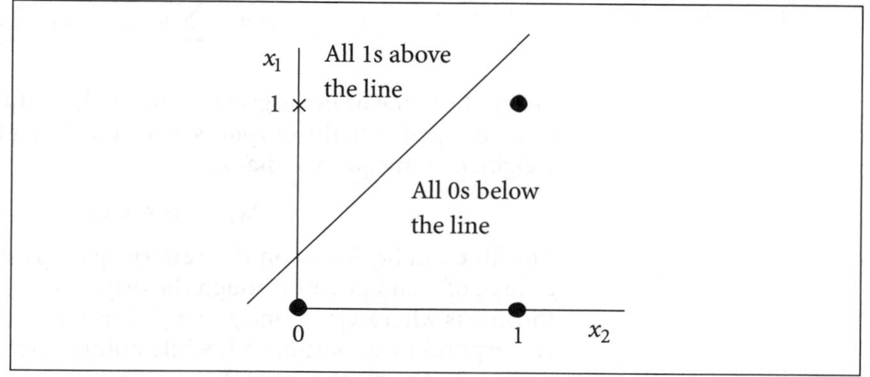

Since a straight line separates the two classes, this function is linearly separable.

Clearly, not all problems have only two inputs. When the number of inputs, n, is more than two, the pattern space is n-dimensional and an n–1 dimensional surface or hyperplane separates the two classes. An example of this could be the three input function $y = (\neg x_2 \wedge x_3) \vee (x_1 \wedge \neg x_2) \vee (x_1 \wedge x_3)$ shown in Figure 3.4 where a two-dimensional surface can be used to separate the 0s from the 1.

Let us now return to the single perceptron. The weighted sum of the inputs is a linear function of those inputs. If its value is greater than 0, the

Figure 3.4 Pattern space showing a hyperplane

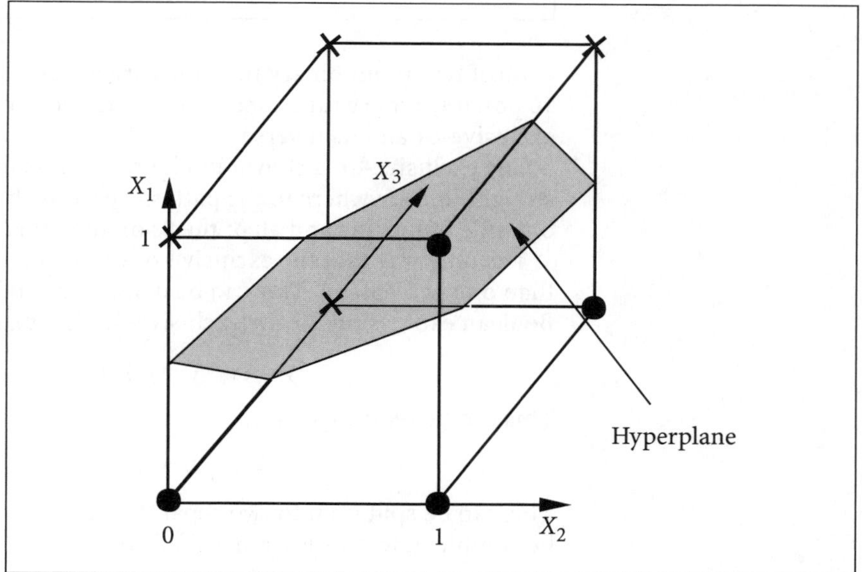

output is 1, or if its value is less than or equal to 0, the output is 0. So the weighted sum separates the two classes of 0 and 1.

The solution that was found earlier for $y = x_1 \wedge \neg x_2$ was 0, 2, –2 for w_0, w_1 and w_2 respectively. The weighted sum, *net*, is therefore:

$$net = \sum_{i=0}^{n} w_i x_i = 2x_1 - 2x_2$$

When the value of *net* is greater than 0, the output is 1, and when it is less than or equal to 0, the output is 0. So a dividing line exists when the weighted sum equals 0, that is:

$$2x_1 - 2x_2 = 0 \text{ or } x_1 = x_2.$$

This line can be drawn on the pattern space as shown in Figure 3.5. It has a slope of 1 and passes through the origin. The only point that lies above the line is where $x_1 = 1$ and $x_2 = 0$. Therefore, points above the line correspond to an output of 1, while points below or on the line correspond to an output of 0.

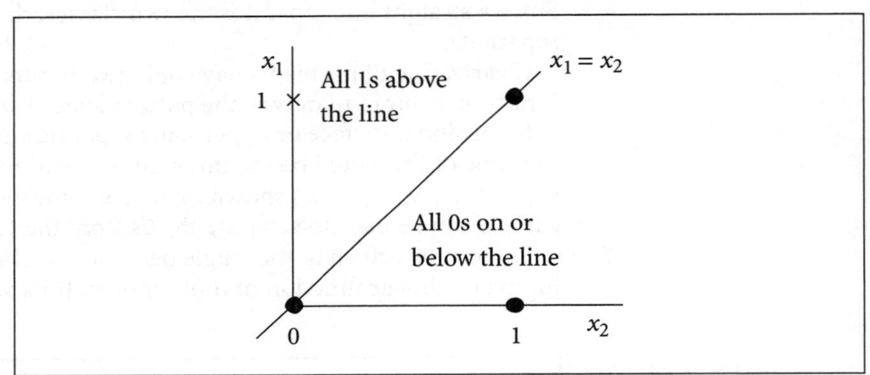

Figure 3.5 Solution to $y = x_1 \wedge \neg x_2$

Most two-input binary functions are linearly separable. In fact, of the 16 possible binary functions only two are not linearly separable – the exclusive-or and its inverse.

The exclusive-or is shown in Figure 3.6. It is impossible to draw a straight line anywhere in the pattern space such that all of the 1s lie on one side of the line and all of the 0s on the other.

The only way that the exclusive-or can be implemented is to use more than one perceptron. This can be demonstrated by rearranging the Boolean expression for the exclusive-or. The usual Boolean expression is:

$$y = (x_1 \wedge \neg x_2) \vee (\neg x_1 \wedge x_2)$$

This can be rearranged as:

$$y = (x_1 \vee x_2) \wedge \neg(x_1 \wedge x_2)$$

This can be split up into two separate functions, y_1 and y_2, which can then be combined to give y again, as follows:

Figure 3.6 The exclusive-or

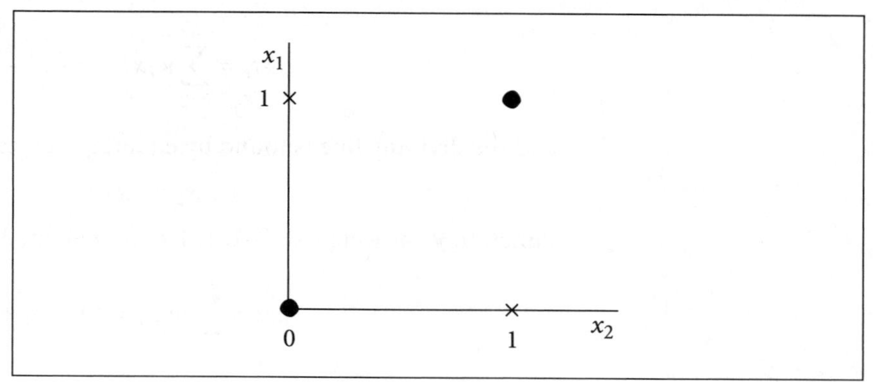

$$y_1 = x_1 \vee x_2$$
$$y_2 = \neg(x_1 \wedge x_2)$$
$$y = y_1 \wedge y_2$$

These are all linearly separable functions, so each one can be implemented using a single perceptron, as shown in Figure 3.7.

Figure 3.7 Two-layer implementation of the exclusive-or

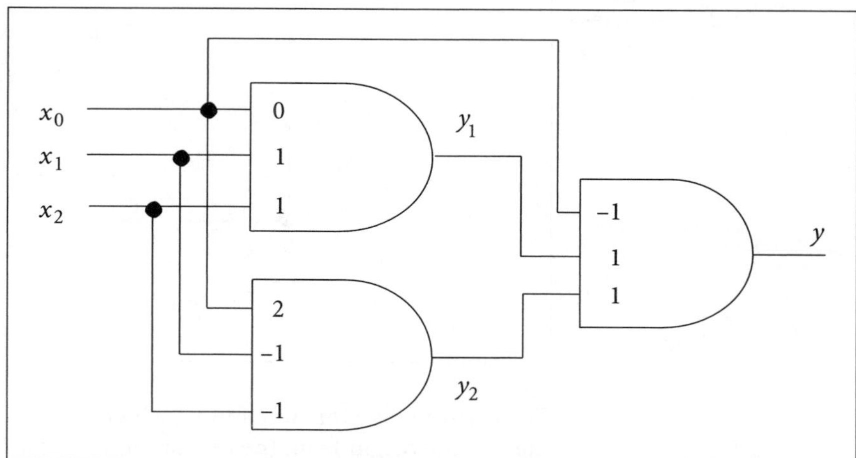

Function y_1 has weights of 0,1,1 and therefore the weighted sum, net_1, is given by the following equation:

$$net_1 = \sum_{i=0}^{n} w_i x_i = x_1 + x_2$$

When the value of net_1 is greater than 0, the output is a 1, whereas if the value of net_1 is less than or equal to 0, the output is 0. So, a dividing line exists at the point when $net_1 = 0$ or:

$$x_1 = -x_2$$

Function y_2 has weights of 2, −1, −1, so the weighted sum, net_2, is:

$$net_2 = \sum_{i=0}^{n} w_i x_i = 2 - x_1 - x_2$$

and the dividing line is found by equating net_2 to 0. This gives:

$$x_1 = -x_2 + 2$$

Function y has weights of –1, 1, 1 so its weighted sum, *net*, is:

$$net = \sum_{i=0}^{n} w_i x_i = -1 + y_1 + y_2$$

and the dividing line has the equation:

$$y_1 = -y_2 + 1$$

The dividing lines for functions y_1 and y_2 can be drawn onto the pattern space, as shown in Figure 3.8. Both have a slope of –1 but have different intercepts on the vertical axis.

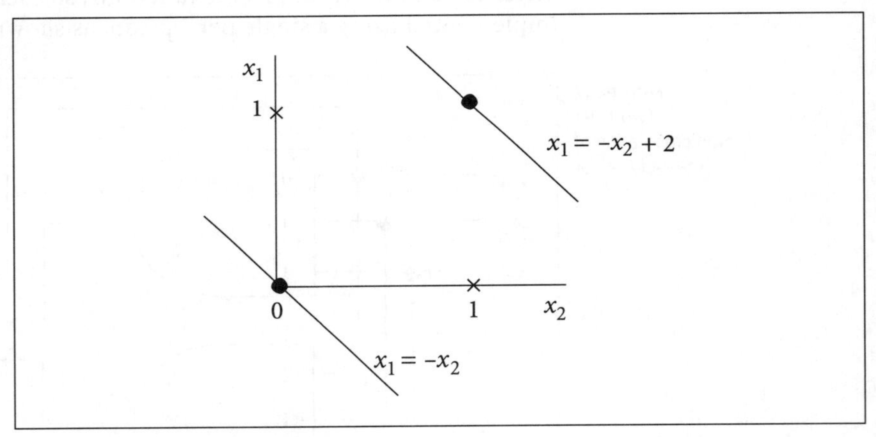

Figure 3.8 The first layer of the exclusive-or

The dividing line for function y_1, which passes through the origin, divides the 0 at the origin from the rest of the 1s and 0. The dividing line for function y_2 divides the 0 at $x_1 = 1$ and $x_2 = 1$ from the other 1s and 0. So the output should be classified as a 1 if it lies above function y_1 and below function y_2. The 0s and the 1s have been successfully separated. This is exactly what function y does, as shown in Figure 3.9.

The dividing line for function y has a slope of –1 and divides the points where y is 0 from those where y is 1. Notice that there are only three points in this figure. That is because the situation where both y_1 and y_2 are 0 cannot occur.

There are other ways of implementing the exclusive-or function using two layers of perceptrons. The point is, however, that it is possible to implement any function, linearly separable or not, by a suitable choice of weights in a multi-layered perceptron. How the weights are chosen is the subject for the next section.

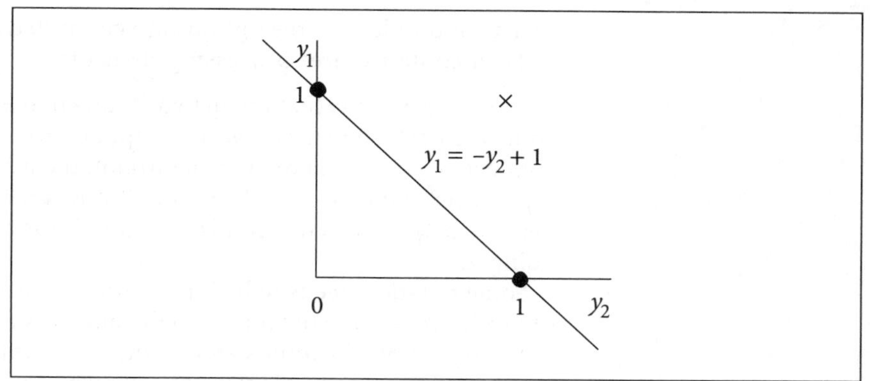

Figure 3.9 Second layer of the exclusive-or

3.3 Back-propagation

We have seen how layers of perceptrons can implement even non-linearly separable functions. It can be shown that all binary functions can be implemented using multi-layered perceptrons, and consequently all pattern classifications. This is quite a grand claim, but do not be misled into thinking that the problem is therefore solved – it is not, but it does at least provide a medium in which to work. If a solution does exist, then with the appropriate choice of the number of layers and the number of perceptrons in each layer it should be possible to find that solution.

Figure 3.10 shows a typical multi-layered perceptron. It has three layers:

- The input layer – the inputs are connected to the outside world. This is basically a 'fan-out' layer where the three inputs are simply connected to the next layer of the network. No processing is done in this layer.
- The hidden layer – this is called the hidden layer because neither the inputs nor the outputs of the two neurons can be seen from the outside. This is the first layer where processing takes place. In a four-layer network there would be two hidden layers.

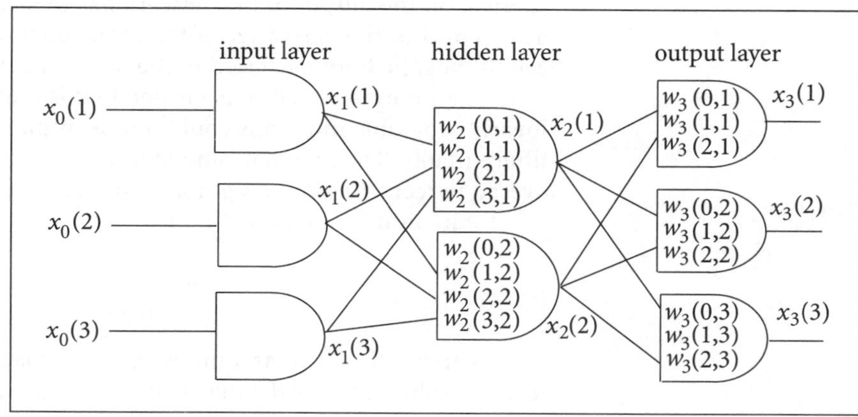

Figure 3.10 A three-layer perceptron

- The output layer – the outputs are connected to the outside world. This layer contains three processing elements.

A further point is that the network shown in Figure 3.10 is fully connected, which means that the output of every neuron in one layer is connected to an input of every neuron in the next layer, starting from the input layer and ending at the output layer. Not every multi-layered perceptron is connected in this way, but this is the most common way of doing it.

Some notation needs to be introduced at this point. Since the outputs of one layer become the inputs to the next it is not sensible to have a different symbol for inputs and outputs, so x will be used to represent the value coming into or going out of a perceptron. Each layer will have a number starting from 1 at the input layer. The inputs to the network are $x_0(0)$, $x_0(1)$, $x_0(2)$, up to $x_0(n)$. The outputs of the first layer, which are the inputs to the second layer are $x_1(0)$, $x_1(1)$, $x_1(2)$ etc. So $x_l(i)$ is the output from the ith neuron in the lth layer.

Next the weights. These will still be called w, but the subscript will denote the layer. So, for example, $w_2(3,1)$ indicates a weight in the second layer (the 2) which is connected to $x_1(3)$, and which contributes to the output $x_2(1)$ via a non-linear function $f(\ )$ as:

$$x_2(1) = f(net_2(1))$$

Note that all weights with a 0 as the first term inside the brackets, $w_l(0,i)$, show the offset for neuron i in layer l.

For fully connected networks, such as the one in Figure 3.10, the value of $net_2(1)$ is:

$$net_2(1) = \sum_{i=0}^{3} w_2(i,1)x_1(i)$$

In order to train a multi-layered perceptron a new learning rule is needed. The original perceptron learning rule cannot be extended to the multi-layered perceptron because the output function, the hard-limiter, is not differentiable. If a weight is to be adjusted anywhere in the network, its effect on the output of the network has to be known, and hence the error. For this, the derivative of the error function with respect to that weight must be found, as used in the original delta rule for the ADALINE.

So a new output function is needed that is non-linear, otherwise non-linearly separable functions could not be implemented, but which is differentiable. The one that is most often used successfully in multi-layered perceptrons is the sigmoid function, shown in Figure 3.11.

The equation for this function is:

$$y = \frac{1}{(1 + e^{-x})}$$

For positive values of x, as x increases y approaches 1. Similarly, for negative values of x, as the magnitude of x increases y approaches 0. In

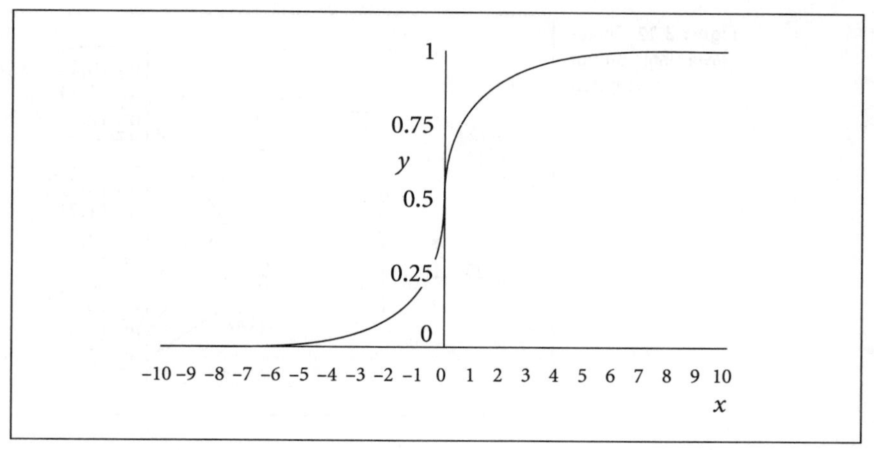

Figure 3.11 The sigmoid function

addition, when $x = 0$, $y = 0.5$. So the output is continuous between 0 and 1 and is therefore differentiable, the derivative being:

$$\frac{dy}{dx} = y' = y(1 - y)$$

The training procedure is similar to the delta rule described in the previous chapter, and is therefore often referred to as the generalised delta rule, although most people call it back-propagation (Rumelhart and McClelland, 1986). The change to a weight w_i in a single perceptron with a sigmoid function is:

$$\Delta w_i = \frac{\eta}{P} \sum_{p=1}^{P} x_{ip} \delta_p$$

The difference between this generalised delta rule and the previous ADALINE delta rule is in the definition of δ_p. In this equation it is defined as $y_p(1 - y_p)(d_p - y_p)$. Also, just as in the original delta rule, this expression is usually approximated to:

$$\Delta w_i = \eta x_i \delta$$

In other words, the weights are updated after each pattern is presented and not after the whole training set is presented. The reason why this is done is because the training set is probably very large, so that the time taken to train becomes intolerable. This has not been shown to be equivalent to minimising the mean squared error, but is widely adopted.

This formula holds for all elements in the multi-layered network. However, the perceptrons in hidden layers which are not directly connected to the output need a different definition for δ.

Take, for example, a weight in the hidden layer of a three-layer perceptron with a single perceptron in the output layer, as shown in Figure 3.12. If the aim is to change $w_2(1,2)$, then using the equations derived in Appendix A, with $p = 2$ as it is the hidden layer, $q = 2$ as it is

Figure 3.12 Three-layer single-output perceptron

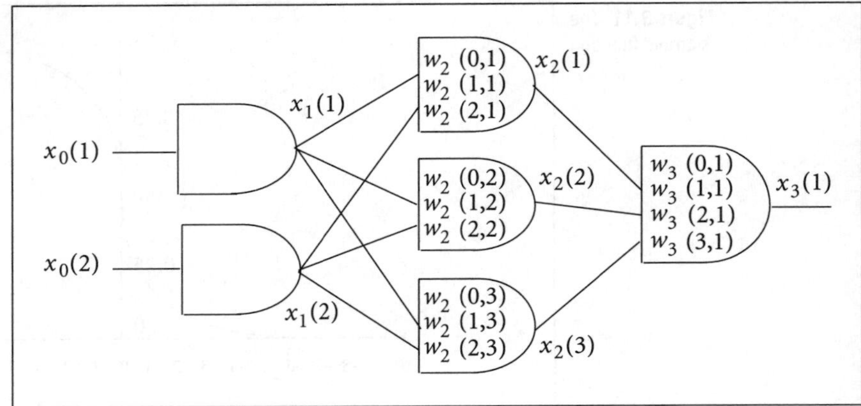

the second neuron in the hidden layer and $r = 1$ as it is the weight connected to the first input, the change is given by the equation:

$$\Delta w_2(1,2) = \eta x_1(1)\delta_2(2)$$

where

$$\delta_2(2) = x_2(2)[1 - x_2(2)]\ w_3(2,1)\delta_3(1)$$

and

$$\delta_3(1) = x_3(1)[1 - x_3(1)][d - x_3(1)]$$

This gives us all the necessary equations that are needed to apply back-propagation to a multi-layered perceptron.

3.3.1 The exclusive-or function

Figure 3.13 shows a network which should be able to implement the exclusive-or function. We have already seen one solution using this network earlier in Figure 3.7.

Figure 3.13 A three-layer network for the exclusive-or

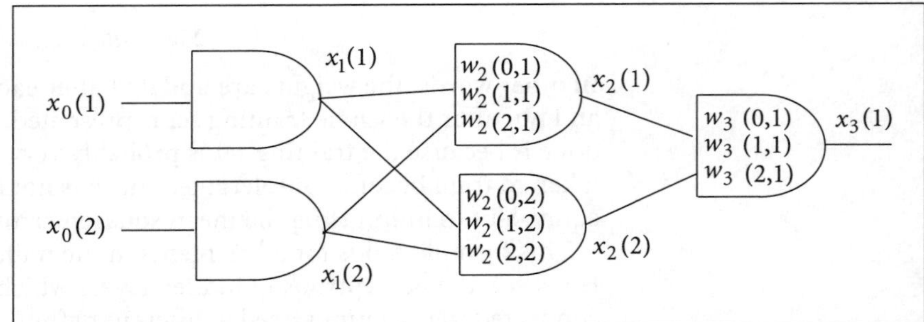

Starting with the following values for the weights, the output of the final neuron can be calculated, and found to be 0.5173481.

$w_2(0,1) = 0.862518$ $w_2(1,1) = -0.155797$ $w_2(2,1) = 0.282885$
$w_2(0,2) = 0.834986$ $w_2(1,2) = -0.505997$ $w_2(2,2) = -0.864449$
$w_3(0,1) = 0.036498$ $w_3(1,1) = -0.430437$ $w_3(2,1) = 0.481210$

Hidden layer	$x_1(0)$	$x_1(1)$	$x_1(2)$	*net*	$x_2(1)$
neuron 1	1	0	0	0.862518	0.7031864
	$x_1(0)$	$x_1(1)$	$x_1(2)$	*net*	$x_2(2)$
neuron 2	1	0	0	0.834986	0.6974081

Output layer	$x_2(0)$	$x_2(1)$	$x_2(2)$	*net*	$x_3(1)$
neuron 1	1	0.7031864	0.6974081	0.0694203	0.5173481

Next, the value of δ for each of the layers is calculated, starting with the output layer.

$$\delta_3(1) = x_3(1)(1 - x_3(1))(d - x_3(1)) = -0.1291812$$

$$\delta_2(1) = x_2(1)(1 - x_2(1))w_3(1,1)\delta_3(1) = 0.0116054$$

$$\delta_2(2) = x_2(2)(1 - x_2(2))w_3(2,1)\delta_3(1) = -0.0131183$$

So the changes to the weights when $\eta = 0.5$ are:

$$\Delta w_2(0,1) = \eta x_1(0)\delta_2(1) = 0.5 \times 1 \times 0.0116054 = 0.0058027$$

$$\Delta w_2(1,1) = \eta x_1(1)\delta_2(1) = 0.5 \times 0 \times 0.0116054 = 0$$

$$\Delta w_2(2,1) = \eta x_1(2)\delta_2(1) = 0.5 \times 0 \times 0.0116054 = 0$$

$$\Delta w_2(0,2) = \eta x_1(0)\delta_2(2) = 0.5 \times 1 \times -0.0131183 = -0.0065592$$

$$\Delta w_2(1,2) = \eta x_1(1)\delta_2(2) = 0.5 \times 0 \times -0.0131183 = 0$$

$$\Delta w_2(2,2) = \eta x_1(2)\delta_2(2) = 0.5 \times 0 \times -0.0131183 = 0$$

$$\Delta w_3(0,1) = \eta x_2(0)\delta_3(1) = 0.5 \times 1 \times -0.1291812 = -0.0645906$$

$$\Delta w_3(1,1) = \eta x_2(1)\delta_3(1) = 0.5 \times 0.7031864 \times -0.1291812 = -0.0454192$$

$$\Delta w_3(2,1) = \eta x_2(2)\delta_3(1) = 0.5 \times 0.6974081 \times -0.1291812 = -0.045046$$

The new values for the weights are therefore:

$w_2(0,1) = 0.868321$ $w_2(1,1) = -0.155797$ $w_2(2,1) = 0.282885$
$w_2(0,2) = 0.828427$ $w_2(1,2) = -0.505997$ $w_2(2,2) = -0.864449$
$w_3(0,1) = -0.028093$ $w_3(1,1) = -0.475856$ $w_3(2,1) = 0.436164$

The network is then presented with the next input pattern and the whole process of calculating the weight adjustment is repeated. This continues until the error between the actual and the desired output is smaller than some specified value, at which point the training stops.

In this example the network converges to a solution. After several thousand iterations the weights are:

$w_2(0,1) = -6.062263$ $w_2(1,1) = -6.072185$ $w_2(2,1) = 2.454509$
$w_2(0,2) = -4.893081$ $w_2(1,2) = -4.894898$ $w_2(2,2) = 7.293063$
$w_3(0,1) = -9.792470$ $w_3(1,1) = 9.484580$ $w_3(2,1) = -4.473972$

With these values the output looks like:

$x_0(0)$	$x_0(1)$	$x_0(2)$	$x_3(1)$
1	0	0	0.017622
1	0	1	0.981504
1	1	0	0.981491
1	1	1	0.022782

This is close enough to what is required. This shows that back-propagation can find a set of weights for the exclusive-or function, provided that the architecture of the network is suitable. The right architecture means one where there are the right number of layers and neurons in the layers for the function to be implemented. The question of how many layers and how many neurons in each layer is still a subject for research, and is discussed in more detail in the next section.

3.3.2 The number of hidden layers

The multi-layered perceptron is able to carry out complex classifications provided that there are a sufficient number of layers of perceptrons in the network with a sufficient number of perceptrons in each layer. A useful guide to the maximum number of layers comes from a theorem put forward in 1957 by the Soviet mathematician A. N. Kolmogorov.

The Kolmogorov Existence Theorem is summarised as follows (Lippmann, 1987).

> This theorem states that any continuous function of n variables can be computed using only linear summations and non-linear but continuously increasing functions of only one variable. It effectively states that a three layer perceptron with $n(2n + 1)$ nodes using continuously increasing non-linearities can compute any continuous function of n variables. A three layer perceptron could then be used to create any continuous likelihood function required in a classifier.

Further support for this conclusion can be found (Lapedes and Farber, 1988) where it is argued that:

> one does not need more than two hidden layers for processing real valued input data, and the accuracy of the approximation is controlled by the number of neurons per layer, and not the number of layers.

However, they do go on to say that two layers of hidden neurons is not necessarily the most efficient number. Using more layers could result in a smaller number of neurons in the whole network.

Research has shown (Hornick *et al.*, 1989) that only three layers are needed. This work gives a theoretical proof that three-layer perceptrons with a sigmoid output function are universal approximators, which means that they can be trained to approximate any mapping between the inputs and the outputs. The accuracy of the approximation depends on the number of neurons in the hidden layer. This is quite a profound quality, and one which has many applications in areas such as control, data fusion and financial forecasting.

3.3.3 The numeral classifier

Let us have another look at the numeral classifier that was described in Chapter 1. There are 64 binary inputs and 10 binary outputs. The main questions that need answering are how many layers are needed and how many neurons in each layer? The problem may be linearly separable, in which case only one layer would be needed. The previous section said that the maximum number of layers required is three but does not say how many neurons are needed in each layer or if three layers is the optimum.

The encoder problem (Ackley *et al.*, 1985) sheds some light on the number of neurons required in each layer. The information about the class the input pattern belongs to has to pass through each layer, so information theory says that if a binary code is used, the number of bits, n, required to code m separate items has to satisfy the following equation:

$$n \geq \log_2(m)$$

In the numeral classifier, $m = 10$, so the smallest whole number that satisfies this formula is $n = 4$. Assuming that the outputs of the neurons in each layer are binary, there has to be at least four neurons in each layer in order for the information to pass through correctly.

Next, there is the value of η. It should be between 0 and 1, but what is a 'good' value to choose?

One solution that implements the numeral classifier problem uses only three layers with 64 neurons in the input layer, and 10 neurons in the hidden layer and the output layer. Since 10 neurons should, in theory, be more than enough to encode the problem, it is not surprising that these values allow the network to successfully train using the training set shown in Figure 3.14. The learning coefficient was set to $\eta = 0.5$.

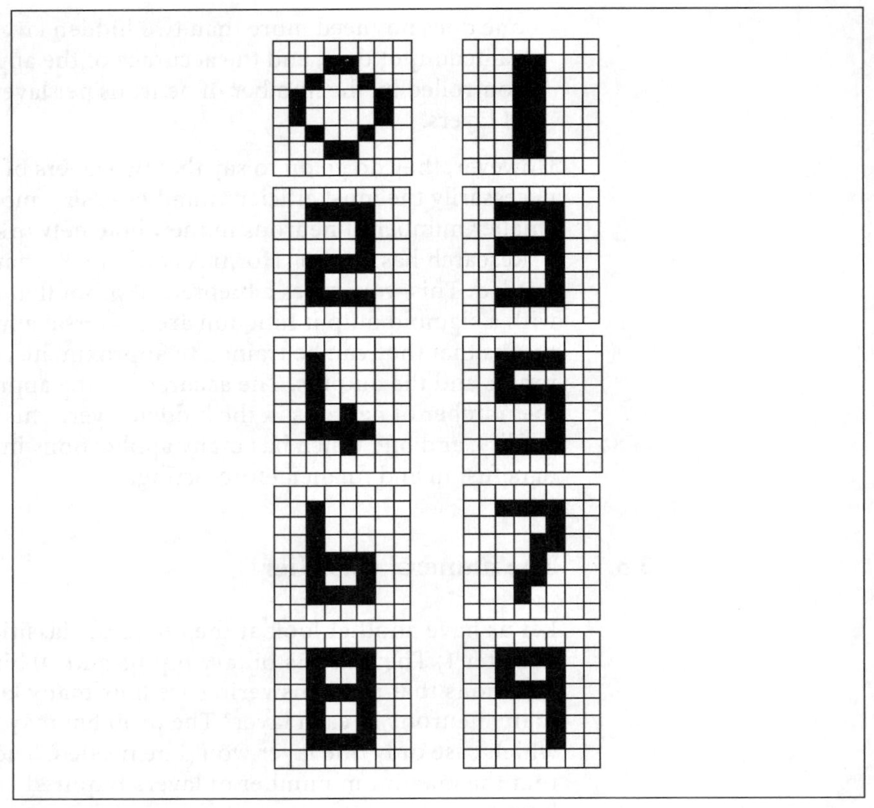

Figure 3.14 Training set for the numeral classifier

3.4 Variations on the standard multi-layered perceptron

Needless to say, although the multi-layered perceptron with one hidden layer and with the sigmoid function as the non-linear output function is probably the most common form, there are many variations. These variations have been introduced usually with the intention of speeding up training.

So far, the changes to weights throughout the perceptron are governed by the equation:

$$\Delta w = \eta x \delta$$

A modification which can sometimes speed up convergence is the addition of a momentum term, which equals a constant, α, times the current Δw value.

$$(\Delta w)_{k+1} = \eta x \delta + \alpha (\Delta w)_k$$

Choosing a suitable value for α is a matter of trial and error, although it is usual to select a value that is small in relation to the value of $\eta x \delta$.

Another variation is to let the output range from –1 to +1 rather than 0 to 1. It is claimed (Haykin, 1999) that this can speed up training because the function is anti-symmetric i.e. $f(-x) = -f(x)$. This, in turn, is more likely to produce output values with zero mean than the standard sigmoid function which produces all positive output values. This range is achieved by adjusting the sigmoid function as follows:

$$y = \frac{2}{1 + e^{-x}} - 1 = \frac{(1 - e^{-x})}{(1 + e^{-x})}$$

If this function, called the hyperbolic tangent, is used then the back-propagation rule has to be modified so that the derivative of y in terms of x is:

$$y' = 2y(1 - y)$$

This means that the value of η, the learning coefficient, can be doubled.

There are many more complex variations which involve varying the learning coefficients, making better estimations of the gradients and how far to move along those gradients, and some which make assumptions about the data to improve learning. Many of these techniques are reviewed in Haykin (Haykin, 1999) and will not be described in detail here.

A final variation is to use trigonometric functions such as $\sin(x)$ and $\tan(x)$ instead of the sigmoid (Baum, 1986). This is claimed to improve the performance of the network when the inputs are signals that could be represented as Fourier series.

3.5 Stopping training

Having discussed how training works and how it can be speeded up, it is worth looking at what happens during training in more detail – in particular the question of how do you know when to stop training?

When training a neural network you start with a set of data. The first step is to split the data into a training set and a test set. You train the network using the training set of data, then you test it using the test set to see how well it copes with previously unseen data.

During training the error is measured between the desired output and the actual output, and the aim is to reduce this error by adjusting the weights. So during training the error (or more precisely the squared error) should drop, and training stops when this reaches a sufficiently low value.

It is tempting to think that you would like the error to be as small as possible, so that the neural network can reproduce the training data as closely as possible. However, experience shows that the network never performs as well on the test set as it did on the training set. This phenomenon is known as over-training, and the network is said to over-fit the training data.

In Chapter 1 it was said that the fundamental properties of a neural network is its ability to learn and to generalise. By over-training the data, the ability to generalise has been reduced. So the question now is how do you know when to stop training so that you avoid over-training?

The solution is to split the data set into three sets rather than two. These sets are usually called the training set, test set and validation set. Training is done using the training set and the error used to adjust the weights. However, during training the network is also shown the test data and the error recorded. Note that when the network is shown the test data the weights are not adjusted – only after being presented with the training data are the weights adjusted using back-propagation.

During this training process the error between the desired output and the actual output should fall. However, the error in the test data should also fall but be higher than the error in the training data, as shown in Figure 3.15. At some point in the training the error in the test data stops falling and could even start to rise. This is where over-training has began, and the network is starting to over-fit the training data. If training is stopped at the point where the error in the test data starts to rise, then over-training can be avoided. Effectively a compromise is reached where the network is trained to give a good performance on the training set and the test set, rather than giving a better performance on the training set at the expense of a worse performance on the test set.

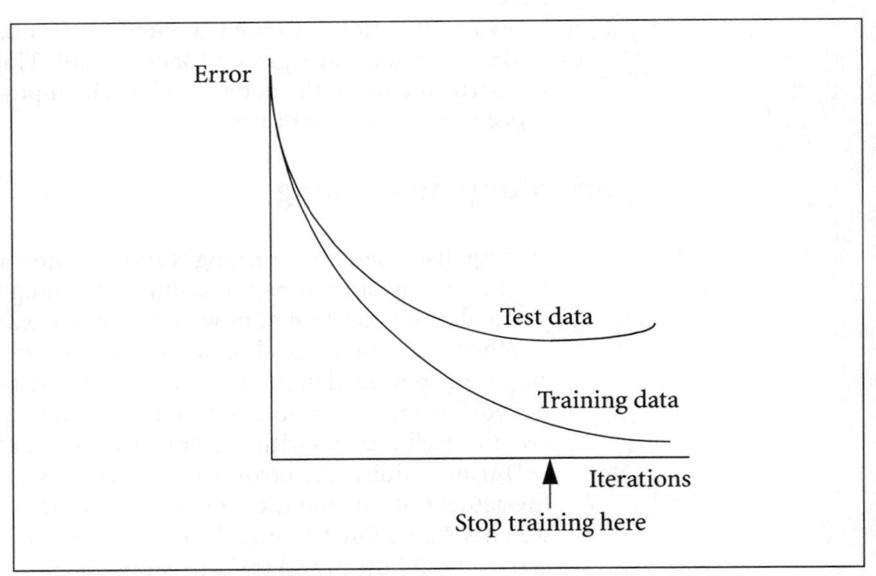

Figure 3.15 Error in the training and test sets during training

Finally, the network is presented with the validation set, and the performance is judged on this set which contains previously unseen data.

CHAPTER SUMMARY

This chapter described the perceptron with its learning rule. The limitations of a single perceptron, of only being able to implement linearly separable functions, can be overcome if multiple layers of perceptrons are used. The perceptron itself and the learning rule then have to be modified so that back-propagation of errors can take place. The modification to the perceptron is the change from a hard-limiter as the non-linear output function to a sigmoid function which is differentiable. The modification to the learning rule is a redefinition of δ for the perceptrons in the hidden layers.

The multi-layered perceptron should be capable of implementing mappings between input patterns and the required output responses. However, there are a number of factors which are not well understood, namely the number of layers required, the number of neurons in each layer and the best values for η and α. In each case there seems to be a balance between too high and too low or too many or too few. A wrong selection can mean that the network takes an inordinately long time to converge onto a solution or, worse still, never converges at all. Despite the convergence theorem for perceptrons, if there are not enough layers or not enough elements in each of the layers the network may not converge.

It has been suggested (Rumelhart and McClelland, 1986) that even if a suitable number of elements and layers are chosen, in other words a solution should exist, sometimes the system will not converge. Even in the case of the simple exclusive-or problem with two inputs, two hidden neurons and one output neuron, the system sometimes fails to converge.

At the beginning of the chapter it was stated that the multi-layered perceptron is the most frequently used neural network. This is due mainly to the fact that it can perform as a functional approximator, implementing a mapping between the inputs and the outputs. As stated later in the chapter, this property means that it is used in such diverse applications as image interpretation (Hopgood *et al*, 1993), path planning in robotics (Meng and Picton, 1992) and control engineering. The latter application will be discussed in more detail in a later chapter.

SELF-TEST QUESTIONS

1 The pattern space of Figure 3.16 contains examples of two features, labelled A and B. Is it possible to find a set of weights in a single perceptron that could successfully separate all of the data from the two features?

2 A single neuron has four inputs (including x_0) with the following weights and input values:

i	x_i	w_i
0	1	0.2
1	0.6	−0.4
2	0.5	−0.1
3	0.2	0.3

What is the value of the output when a sigmoid function is used as the non-linear output function?

3 A multi-layered perceptron has two input units, two hidden units and one output unit. What are the formulae for updating the weights in each of the units?

Note: remember to include the weights, w_0, for the bias in each neuron.

4 What is the value of the output of the network described in the previous question if the weights are given by the following table and both inputs are +1.0?

$w_2(0,1)$	$w_2(1,1)$	$w_2(2,1)$	$w_2(0,2)$	$w_2(1,2)$	$w_2(2,2)$	$w_3(0,1)$	$w_3(1,1)$	$w_3(2,1)$
1.7	2.6	0.2	−0.1	0.7	1.5	0.5	1.2	−0.3

What are the values of δ for the same network with these inputs if the desired output is +1.0?

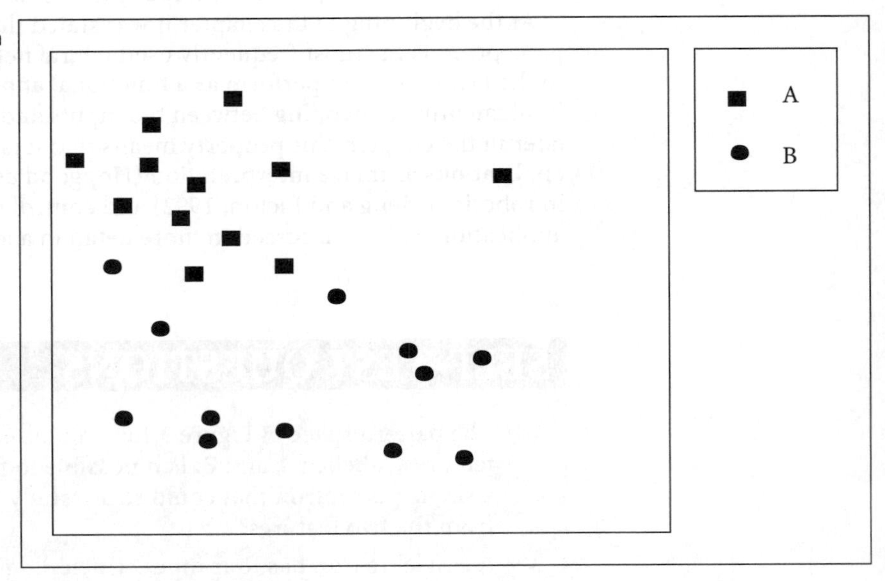

Figure 3.16 Data for Question 1

SELF-TEST ANSWERS

1 No. It is impossible to draw a straight line on the diagram which would separate the two groups. You need at least two lines.

2 $net = 1 \times 0.2 + 0.6 \times (-0.4) + 0.5 \times (-0.1) + 0.2 \times 0.3 = -0.03$

$$y = 1/(1+e^{-net}) = 1/(1+1.03) = 1/2.03=0.493$$

3 For the unit in the output layer:

$$\Delta w_3(0,1) = \eta\delta_3(1)x_2(0)$$

$$\Delta w_3\ (1,1) = \eta\delta_3(1)x_2(1)$$

$$\Delta w_3\ (2,1) = \eta\delta_3(1)x_2(2)$$

where $\delta_3(1) = x_3(1)[1 - x_3\ (1)][d_1 - x_3(1)]$.
For the first unit in the hidden layer:

$$\Delta w_2(0,1) = \eta\delta_2(1)x_1(0)$$

$$\Delta w_2(1,1) = \eta\delta_2(1)x_1(1)$$

$$\Delta w_2(2,1) = \eta\delta_2(1)x_1(2)$$

where $\delta_2(1) = x_2(1)[1 - x_2\ (1)]w_3(1,1)\delta_3(1)$.
For the second unit in the hidden layer:

$$\Delta w_2(0,2) = \eta\delta_2(2)x_1(0)$$

$$\Delta w_2(1,2) = \eta\delta_2(2)x_1(1)$$

$$\Delta w_2(2,2) = \eta\delta_2(2)x_1(2)$$

where $\delta_2(2)= x_2(2)[1 - x_2(2)]w_3(2,1)\delta_3(1)$.

4 To calculate the output of the network start at the hidden layer and work forward. For the first neuron in the hidden layer:

$$net = 4.5 \qquad x_2(1) = 0.989$$

For the second neuron in the hidden layer:

$$net = 2.1 \qquad x_2(2) = 0.891$$

For the only neuron in the output layer:

$$net = 1.142 \qquad x_3(1) = 0.805$$

To calculate the values of δ, start at the output and work backwards. At the only neuron in the output layer:

$$\delta_3(1) = x_3(1)(1 - x_3(1))(d - x_3(1))$$
$$= 0.805(1 - 0.805)(1 - 0.805)=0.031$$

At the hidden layer, the first value is:

$$\delta_2(1) = x_2(1)(1 - x_2(1))w_3(1,1)\ \delta_3(1)$$
$$= 0.989(1 - 0.989)(1.2)(0.031) = 0.0004$$

and the second one is:

$$\delta_2(2) = x_2(2)(1 - x_2(2))w_3(2,1)\ \delta_3(1)$$
$$= 0.891(1 - 0.891)(\ -0.3)(0.031) = -0.0009$$

CHAPTER 4

Boolean neural networks

CHAPTER OVERVIEW

This chapter gives a broad introduction to Boolean neural networks, particularly the specific device known as the WISARD. It examines:

- examples of Boolean networks
- how a Boolean neural network learns and generalises
- the performance of the WISARD system
- how a logic node works
- how probabilistic logic nodes work

4.1 Bledsoe and Browning's program

Boolean neural networks differ from networks such as the ADALINE and perceptron by using Boolean logic elements as the basic components. The simplest Boolean logic elements are logic gates such as the AND gate. However, more complex devices such as memories can also be described as Boolean logic devices since they could be built from these simpler gates. This means that the basic action of the elements is logical rather than involving the summation of weighted inputs. This gives rise to another term which is sometimes used to describe them, namely weightless neural networks.

The WISARD that was described briefly in Chapter 1 is an example of a Boolean neural network which is built using RAM chips. We shall return to the WISARD shortly, but first we will look at an earlier network (Bledsoe and Browning, 1959).

Let us return to the problem that was described in Chapter 1 – the numeral classifier. The input is an 8×8 binary image, so there are 64 pixels in all. There are 10 binary valued outputs which indicate which numeral is at the input. An output of all 0s indicates that the input is not a numeral.

A solution that is based on Bledsoe and Browning's program is shown in Figure 4.1. The 64 pixels are randomly connected in pairs to the address lines of 32 four-word memories, where each of the words are 10

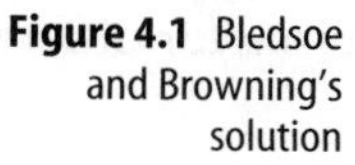

Figure 4.1 Bledsoe and Browning's solution

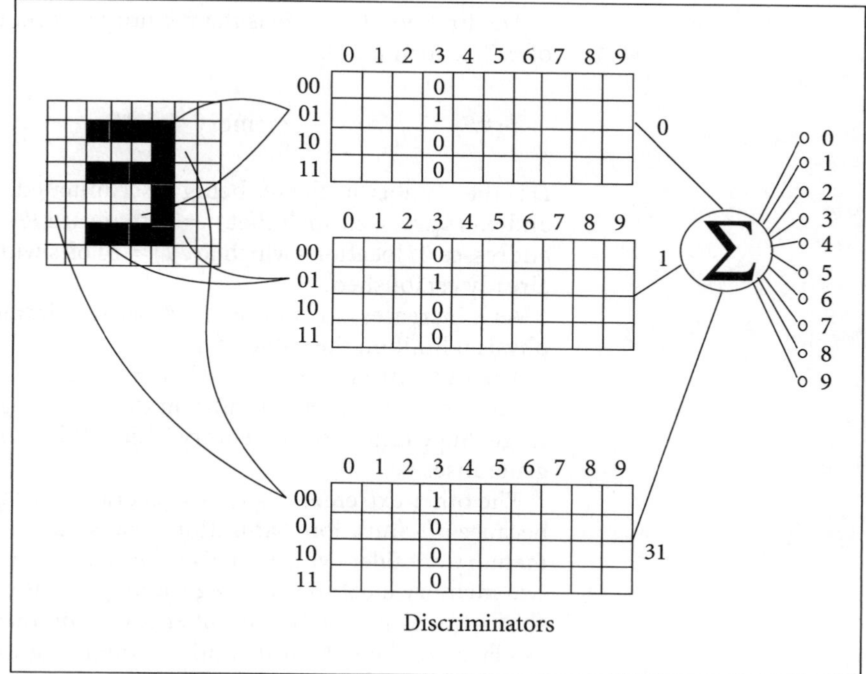

bits wide. Since there are 32 four-word memories, all of the pixels are connected to the memory. The reason for randomly connecting the pixel pairs will be discussed later.

The pairs of pixels can have any of four values – 00, 01, 10, or 11. These are the addresses of each of the four 10-bit words. During training, the system is shown several examples of each of the numbers. For any one particular example, all 32 of the memories have a 1 placed in them at the location determined first by which of the four addresses are formed by the pixel pairs and secondly by which of the numbers is being shown.

In Figure 4.1, the number 3 is being shown. Some of the links are shown to the memories. For example, memory 0 has the address 01 applied, so a 1 is placed at line 1, bit 3. Similarly, all the other 31 memories have a 1 placed in bit 3 at one of the four words.

After training, a number is shown at the input, and the outputs of the 32 memories are summed. The 10 outputs can range in value from 0 to 32, the one finally selected being the one with the highest total. Ties are resolved randomly.

The memory size for this particular problem is surprisingly small. In all there are 32 four-word memories, where each word is 10 bits wide. The total is therefore 1280 bits.

A question which Bledsoe and Browning asked in their original paper was what happens if triplets or groups of n pixels are used instead of just pairs? They called these groups n-tuples, where n can have a value of anything between 1 and P, where P is the total number of pixels.

The first point to note is the memory size required for different values of n. The formula is:

$$\text{memory size } M = \frac{P \times 2^n \times D}{n}$$

D is the number of classes being discriminated. In the case where $P = 64$, and pairs are used such that $n = 2$, there are $P/n = 32$ pairs. Each pair addresses 2^n locations which are $D = 10$ bits wide. So $M = 1280$ bits, as already established.

As n increases, so does the memory requirements, so this imposes a physical limit on the value of n.

The two extreme cases are when $n = 1$ and $n = P$, the latter being equivalent to the original solution that was suggested in Chapter 1 using a single huge impractical memory which did not have the ability to generalise.

The other extreme, $n = 1$, can generalise but quickly becomes useless because of saturation. Saturation means that all of the locations in the memory are filled with 1s. This is because each pixel addresses either of two memory locations. During training, let us suppose that it is being shown an example of the number 3. Certain memory locations will have 1s placed in them. Now if another example of the same number is shown, only slightly different, such as by being shifted one pixel to the left say, then a large number of pixels which were 1 are now 0 and vice versa. Consequently the memory locations connected to those pixels will now have a 1 placed at the alternative address. That discriminator is now useless, since it has a 1 in both addresses.

When a larger value of n is chosen, saturation becomes less of a problem. However, for very large values of n, the system loses its ability to generalise. There is a trade-off, therefore, between systems with small values of n saturating, and systems with large values of n not being able to generalise and not being physically realisable.

A study has been done (Ullmann, 1969) that shows the relationship between the recognition performance and the value of n. Typical curves are shown in Figure 4.2 for a numeral classifier which was trained using handwritten examples of each numeral digitized into 22 × 30 pixels.

It can be seen that for a particular sample size, as the value of n increases the percentage of correct classifications increases until a peak is reached. After that, as n increases the percentage of correct classifications falls off to zero. If the sample size increases, a higher peak can be achieved with some larger value of n, but the shape of the curve remains the same.

The conclusion that can be drawn is that there is a balance to be made between the memory size, the amount of training data and the value of n. Ullmann concluded that:

> however big the training set, there will always in practice be a pool of rarely occurring patterns which the system will fail to recognise correctly.

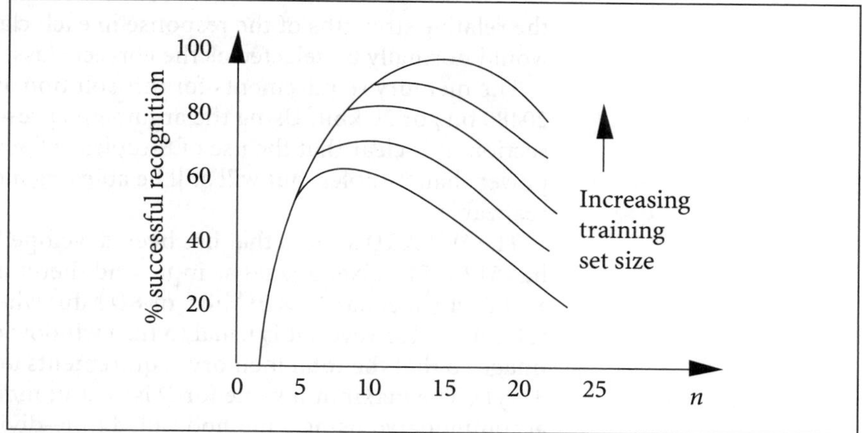

Figure 4.2 Recognition performance versus *n*

He goes on to say that this is a weakness of the *n*-tuple method, but this is probably also a weakness in all other recognition systems.

4.2 WISARD

The WISARD is an extension of Bledsoe and Browning's machine. Instead of randomly connecting pairs of pixels to the address lines of the memories, more pixels are used, usually 8. So each separate memory, called a discriminator, has 256 words.

Figure 4.3 shows the WISARD solution to the same numeral classification problem. Training and operation are the same as Bledsoe and Browning's program. This time there are 64/8 = 8 memories. Notice that the output is usually displayed as a bar chart on a monitor showing

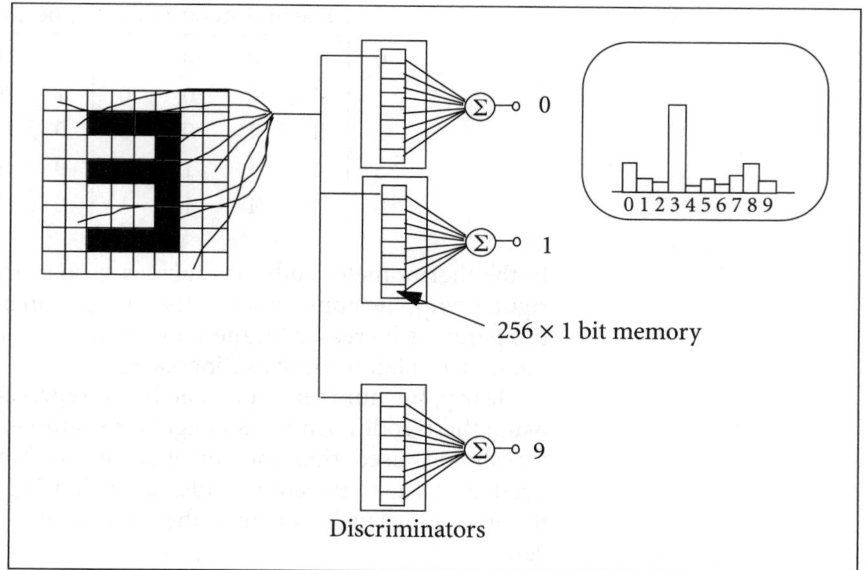

Figure 4.3 WISARD solution to the number classification problem

the relative strengths of the response in each class. The largest response would normally be selected as the correct class.

The memory requirements for this solution are $8 \times 8 \times 2^8 \times 10/8 =$ 20480 bits or 20 kbit. Using the arguments presented in the previous section, it is clear that the use of 8-tuples will give greater discriminating power than 2-tuples, but will still be able to generalise and be physically realizable.

The WISARD system that has been developed (Aleksander *et al.*, 1984) has 512×512 pixel images as input and therefore requires $512 \times 512 \times 2^8 \times D/8$ which equals $D \times 2^{23}$ bits, or $8D$ Mbit when using 8-tuples over the full image. However, it is usual to fix a window of dimensions x,y over the image so that the total memory requirements are only $x \times y \times 2^8 \times D/8 = 32xyD$. The maximum value for D is 16, but more classes can be accommodated using a method called time-division multiplexing.

The great advantage of the system is that all of the 'computation' is done in parallel, so that the system can operate in real time, which usually means 25 pictures or frames per second.

4.2.1 Encoding grey levels

The images that are used in the WISARD are 8-bit grey level, ranging from 0 to 255. The WISARD, like all other Boolean neural networks, can only handle binary data, so the input has to be coded. In many applications it is sufficient to threshold the grey scale image to produce a binary image. This means that all values of grey level above a certain threshold become white and all values below the threshold become black.

Some often used alternatives are thermometer codes (Aleksander and Wilson, 1985) and 1-in-n codes (Aleksander and Stonham, 1979). These can be demonstrated using a 4-bit binary number as follows:

i	Thermometer code				One-in-4 code			
0	0	0	0	0	0	0	0	0
1	1	0	0	0	1	0	0	0
2	1	1	0	0	0	1	0	0
3	1	1	1	0	0	0	1	0
4	1	1	1	1	0	0	0	1

In the thermometer code, as i increases, so more 1s are added from left to right, giving the appearance of the mercury in a thermometer rising as the temperature increases. In the 1-in-4 code, only one bit is ever 1, and it moves from left to right as i increases.

Clearly, the number of bits needed to represent each pixel increases using these codes. An 8×8 image, for example, has 64 pixels. If a threshold is used, then each pixel becomes a binary value, so 64 bits are needed. If one of the above codes is used, using four bits to represent different grey levels, say, then the total number of bits required is $4 \times 64 =$ 256.

Some novel ways of coding grey-scale n-tuples which are beyond the scope of this book are rank ordering (Austin, 1988) or Minchinton cells (Minchinton *et al*, 1990).

4.2.2 Some analysis of the WISARD

When the WISARD is training it is shown a pattern and each of the K RAMs in one particular discriminator is addressed by an n-tuple and 1s are written into the locations that are addressed. From the discussions in the previous sections, if the picture size is P pixels, then the value of K is:

$$K = \frac{P}{n}$$

Later, when the WISARD is shown a new pattern, there will be an overlap between that pattern and the original pattern that it was shown during training. Overlap here means that the pixel values are the same, and it is assumed that the image has already been converted to a binary image. Let the overlapping area, that is, the number of pixels with the same value, be A.

For a particular RAM in the jth discriminator, the probability of firing depends on the probability of the n address lines having identical values for both patterns. This occurs when the n-tuple is taken from overlapping areas. The probability of selecting n pixels from overlapping areas is equal to:

$$p = \left(\frac{A}{P}\right)^n$$

If the value of K is large, then it is probable that p of the K RAMs in the discriminator will fire. The value of the output of the jth discriminator, r_j, assuming that the outputs of the RAMs are summed, will therefore be:

$$r_j = K\left(\frac{A}{P}\right)^n$$

Clearly, if the two patterns are identical then $A = P$ and the output $r_j = K$. If this output is 1, that is, the responses are normalized, then the normalized value of r_j is:

$$r_j = \left(\frac{A}{P}\right)^n$$

Although this value has been derived using a very simple situation where there was only one pattern, it is approximately the same for more complex situations (Aleksander, 1983).

A further measure which can be calculated is the confidence, C, with which a decision is made. The expression for C is:

$$C = \frac{r_1 - r_2}{r_1} = 1 - \frac{r_2}{r_1}$$

where r_1 is the response of the strongest discriminator and r_2 is the response from the strongest incorrect discriminator. Complete confidence, $C = 1$, can only be achieved when $r_2 = 0$ which happens when only one discriminator responds. This is unlikely to happen, as it corresponds to a situation where there is no overlap between the test pattern and any training pattern from other classes.

Substituting for r gives:

$$C = 1 - \left(\frac{A_2}{A_1}\right)^n$$

The importance of this equation is that it shows how the confidence improves as n increases. Since, by definition, $A_2 \leq A_1$ it follows that A_2/A_1 is less than 1 and gets smaller as n increases, so C gets closer to 1.

This gives numerical support to the argument presented earlier that as n increases the discriminatory powers of the network improves, and so the confidence in its ability to discriminate also improves. The penalty for increasing n is the loss in the ability to generalise and the increase in size.

4.3 The arguments in favour of random connections

A number of explanations can be found for randomly connecting the pixels to the RAMs. Probably the easiest way is to repeat the example that can be found in Aleksander's book (Aleksander and Burnett, 1984).

Take the 8 × 8 image again, and design a system to be able to discriminate between horizontal and vertical bars. Using 4-tuples, there will be 16 RAMs, each with 16 2-bit entries as shown in Figure 4.4.

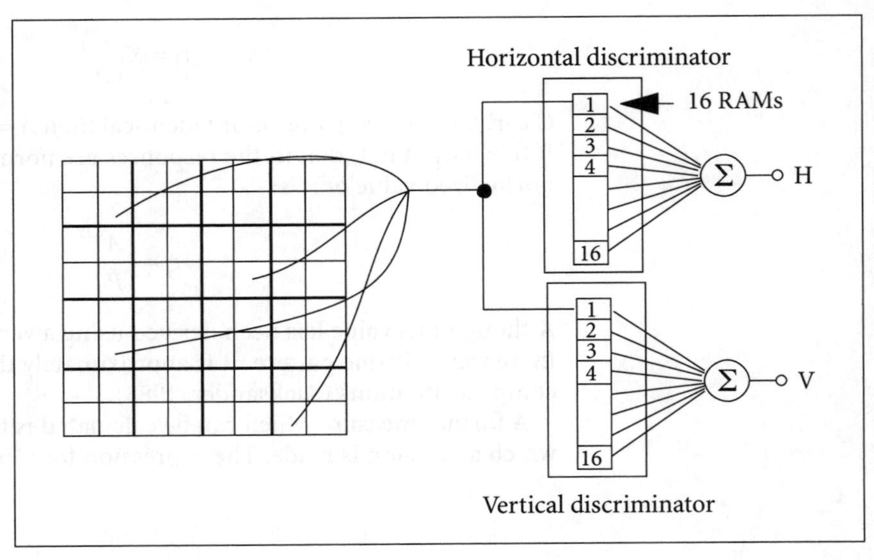

Figure 4.4 The horizontal and vertical bar detector

The system is trained by showing it four non-overlapping 2-pixel wide horizontal and vertical bars, as shown in Figure 4.5.

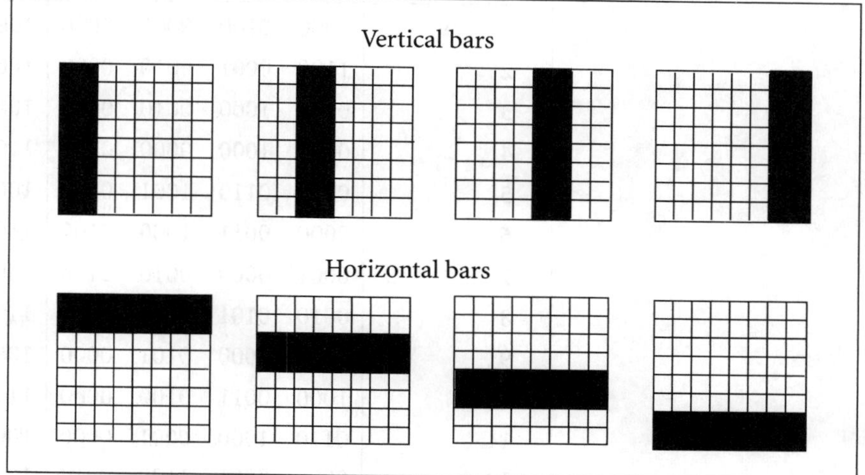

Figure 4.5 Training set

First let us see what happens if the RAMs are connected in a regular way, as shown in Figure 4.6 (a). The numbers in the squares correspond to the RAM to which the block of four pixels are connected. The result of this training is that whenever one of the vertical or horizontal bars from the training set is presented to the system, the output of the vertical *and* horizontal discriminators is 16, the maximum, which shows that the WISARD cannot discriminate between horizontal and vertical bars. If similar 2-pixel wide horizontal bars are presented to the system, which are one pixel displaced from the original training set, the response of both discriminators is 0.

Now, let us see what happens if the system is randomly connected. Figure 4.6(b) shows the 8 × 8 image with the numbers representing the

Figure 4.6 (a) Regularly connected RAMs; (b) Randomly connected RAMs

(a) RAM

1		2		3		4	
5		6		7		8	
9		10		11		12	
13		14		15		16	

(b) RAM

1	2	3	4	5	6	7	8
9	10	11	8	12	10	6	4
11	13	6	10	14	9	15	13
2	8	1	5	2	12	16	7
3	9	6	10	7	13	12	1
11	7	15	5	9	3	15	4
16	14	2	8	1	16	14	16
13	3	15	12	5	11	4	14

Table 4.1 Addresses where 1s are stored in each of the RAMs

Number	Vertical discriminator bar 1	bar 2	bar 3	bar 4	Horizontal discriminator bar 1	bar 2	bar 3	bar 4
1	1000	0100	0001	0010	1000	0100	0010	0001
2	1100	0001	0010	0000	1000	0110	0000	0001
3	0101	1000	0010	0000	1000	0000	0110	0001
4	0000	1000	0000	0111	1100	0000	0010	0001
5	0000	0110	1001	0000	1000	0100	0010	0001
6	0000	0011	1000	0100	1100	0010	0001	0000
7	0001	0000	0010	1100	1000	0100	0011	0000
8	0010	0101	0000	1000	1100	0010	0000	0001
9	1010	0000	0101	0000	1000	0100	0011	0000
10	1000	0011	0100	0000	1100	0010	0001	0000
11	0110	1000	0001	0000	1000	0100	0010	0001
12	0000	0001	1100	0010	1000	0110	0010	0001
13	1001	0000	0010	0100	0000	1100	0010	0001
14	0100	0000	1000	0011	0000	1000	0000	0111
15	0000	0011	1000	1100	0000	1000	0110	0001
16	0100	0000	0010	1001	0000	1000	0000	0111

RAM number to which each pixel is connected. So, for example, RAM number 1 is connected to four pixels, distributed all over the image.

Table 4.1 shows the addresses where a 1 is stored in each of the 16 RAMs after training with the same set of horizontal and vertical bars. Table 4.2 shows the scores that are found when the examples from the training set are shown again. Clearly, the system is now able to discriminate between them.

Although this horizontal and vertical bar detector showed that random connections can recognise the training data, it needs to be tested on unseen data. Look at what happens when a vertical bar is shown which is like one of the training set but displaced by one pixel (Figure 4.7).

The resulting score from the discriminators, shown in Table 4.3, is very much in favour of the pattern being a vertical bar. This example shows that it is possible for a randomly connected system to perform better than an organised system in some circumstances. The reason why the system which was connected in an orderly fashion failed was because the patterns being discriminated were very similar to the way the system was organised, so the system had a kind of 'blind spot'.

If a general pattern classification system is needed, then these blind spots should be avoided, and one way that this can be done is to use

Table 4.2 Scores for the examples in the training set

		Vertical discriminator	Horizontal discriminator
Vertical	Bar 1	16	5
	Bar 2	16	10
	Bar 3	16	7
	Bar 4	16	6
Horizontal	Bar 1	7	16
	Bar 2	6	16
	Bar 3	7	16
	Bar 4	7	16

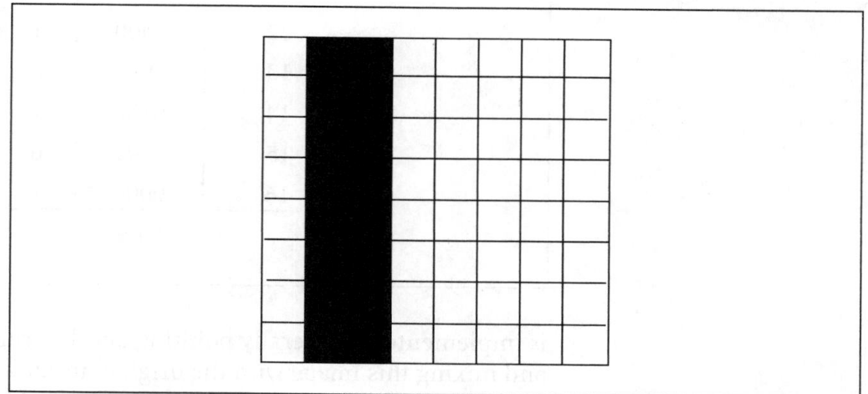

Figure 4.7 A displaced vertical bar

randomness, since, by definition, randomness does not have a pattern to it and cannot therefore overlap with the problem.

4.4 Other work

The description of the WISARD and Boolean networks in general has been restricted to what could be called established work. However, the field of Boolean networks is still being actively researched. The following topics have been included as a sample of some of the work that is being done, which could be the way forward for Boolean networks.

4.4.1 Feedback

Work has been done to extend the capabilities of the WISARD by adding feedback (Aleksander, 1983; Aleksander and Burnett, 1984). The output of a WISARD is usually a bar graph which shows the confidence levels for each of the classes taken from the outputs of the discriminators. Feedback

Table 4.3 Result of showing the WISARD the new vertical bar

RAM	Address	V	H
1	0100	1	1
2	1001	0	0
3	1001	0	0
4	0000	1	1
5	0000	1	0
6	0011	1	0
7	0001	1	0
8	0010	1	1
9	0010	0	0
10	1000	1	0
11	1100	1	1
12	0000	1	0
13	1000	0	0
14	0100	1	0
15	0101	0	0
16	0000	1	1
	Total	11	5

is implemented by literally pointing another camera at this visual display and mixing this image with the original input image to produce a composite input pattern, as shown in Figure 4.8.

It is claimed that feedback improves the confidence levels during a classification. A property which also emerges from this architecture is the presence of a short term memory – the system response persists even after the initial image is removed.

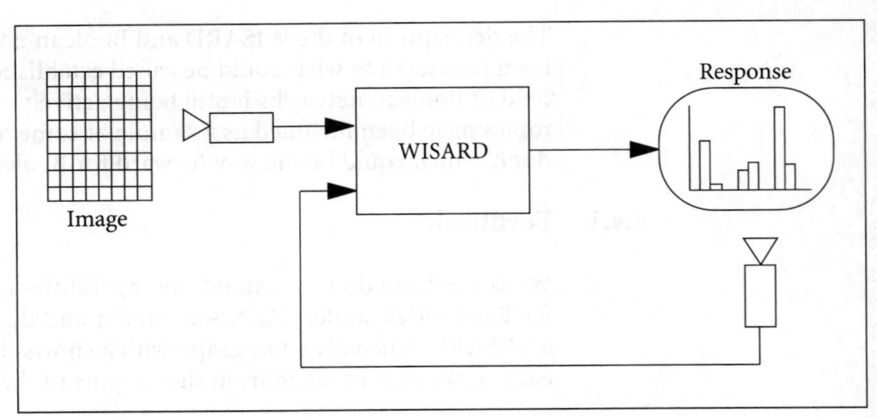

Figure 4.8 WISARD with feedback

4.4.2 Noisy training

Another extension of the WISARD is where the training is done by showing the system good examples of patterns, and letting the system itself extend the training set to include all other patterns that are only one bit different from the good pattern (Aleksander, 1989a). In other words, if the system is shown a good image of the number 3, for example, it will also store 1s in the memory locations addressed by the same input pattern but with one bit inverted so that 'noisy' examples of the number 3 are included in the training set.

An alternative strategy is to add random noise to the good examples during training. So again, good examples are shown, and noisy versions generated from these examples. Both of these methods improve the performance of the WISARD with little extra cost in terms of training time.

4.4.3 Logic nodes

So far we have seen how large pattern classification systems can be built using RAMs. Another possibility is simply to mimic the structures of perceptron-like networks, except replacing the neurons with RAMs. Networks like this are said to consist of logic nodes where each node is in fact a RAM and therefore behaves in a completely logical way (Aleksander, 1989b).

The advantage that a logic node gives is that it can produce an output which is any Boolean function of the inputs. Unlike the perceptron-like neurons it can therefore perform non-linearly separable functions. It can also be implemented using existing technology and analysed using conventional logic techniques.

The major area where these networks are being applied is in feedback networks, performing functions similar to those of the Hopfield and Boltzmann networks which will be discussed in the next chapters. This discussion will therefore be postponed for the moment.

An alternative architecture has recently been proposed that is more like the structure of a multi-layered perceptron but which uses logic nodes (James, 1990). Figure 4.9 shows how the network is used as a character recognition system. The input is a 7×9 binary matrix, so there are 63 pixels which are connected to the address inputs of six $2K \times 8$ bit RAMs. The numbers in the diagram show the number of address or data lines that are connected between RAMs.

Every node in the network is a $2K \times 8$ bit RAM, which therefore has 11 address inputs and 8 data outputs. The data outputs of each layer are connected to the address inputs of the layer in front and in some cases the layer in front of that again. Since there are more address input lines than data output lines, the data get 'condensed' at each layer. It is suggested that the output of the final layer is the 8-bit ASCII code for the character at the input plus information about position, type face, size or any other relevant information.

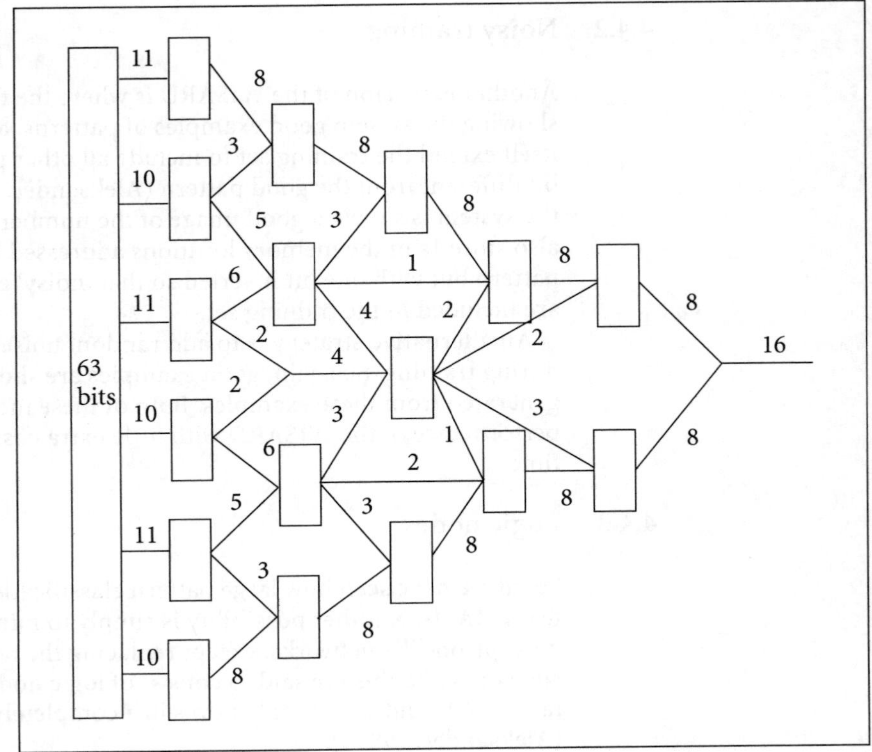

Figure 4.9 James's network

The RAMs are all initially programmed with the numbers 0 to 255, with the sequence repeating eight times. So there are eight possible input patterns on a RAM input that produce the same output. This is how the ability to generalise is built into the system. To train the system, it is shown an input pattern, and the input data is used to address the RAMs in the first layer. The contents of the RAMs are fed forward to the next layer where they are used as address inputs again. This continues until the final layer is reached. In the final layer, the inputs address certain locations in the RAM where the binary or ASCII code for the input pattern can be stored.

Since the contents of all of the RAMs are known, it is possible to work out which input patterns will produce particular outputs. Some of these input patterns will be nonsensical. It is therefore suggested that these unwanted patterns can be eliminated by 'back-propagating' the desired output patterns.

4.4.4 Probabilistic networks

A final variation on Boolean networks is the PLN or probabilistic logic node (Aleksander, 1989a,b) and the P-RAM (Gorse and Taylor, 1989). The standard RAM logic node is a Boolean logic node, BLN, which can only store a 0 or a 1, whereas the PLN stores the probability of the node firing (output a 1).

The device itself is still a RAM which, of course, can still only store 0s and 1s, so the probabilities are derived by interpreting the q-bit number stored in the RAM at each location as a fraction. For example, if q is 8, then the number 00110100 (52) is interpreted as the probability $52/255 = 0.204$.

As in the previous section, PLNs are mainly applied to feedback networks, so these will be discussed further in the next chapters. However, it is clear that they have great potential by introducing probabilistic outputs using conventional technology.

CHAPTER SUMMARY

This chapter has described an alternative approach to the design of pattern classifiers based on Boolean logic rather than being inspired by biological neurons. These Boolean networks behave in similar ways to neural networks, not only because they classify patterns, but because they also learn from training data and generalise.

The great advantage of Boolean networks is the ease with which networks can be implemented using conventional hardware. Boolean networks are also free from the restriction of only being able to implement linearly separable functions – they can implement any Boolean function, so there is no need to have multiple layers. Given all of these advantages, it is hard to understand why Boolean networks have had such a limited impact on the general field of neural networks.

SELF-TEST QUESTIONS

1. What are the memory requirements of a WISARD system that uses a 16×16 image window and has eight discriminators? You should remember that the WISARD uses 8-tuples.
2. How many different values can be encoded with an n-bit thermometer code?
3. A WISARD system has an 8×8 pixel input and is connected using 4-tuples connected randomly as in Figure 4.6(b). After training with the four horizontal and vertical bars, what would the output of the horizontal and vertical discriminators be if the input was as shown in Figure 4.10?

SELF-TEST ANSWERS

1. The formula used to calculate this is $32xyD$, where x and y are the dimensions of the window and D is the number of discriminators. Therefore the memory requirements are $32 \times 16 \times 16 \times 8 = 65\,536$ bits or 64 kbit.

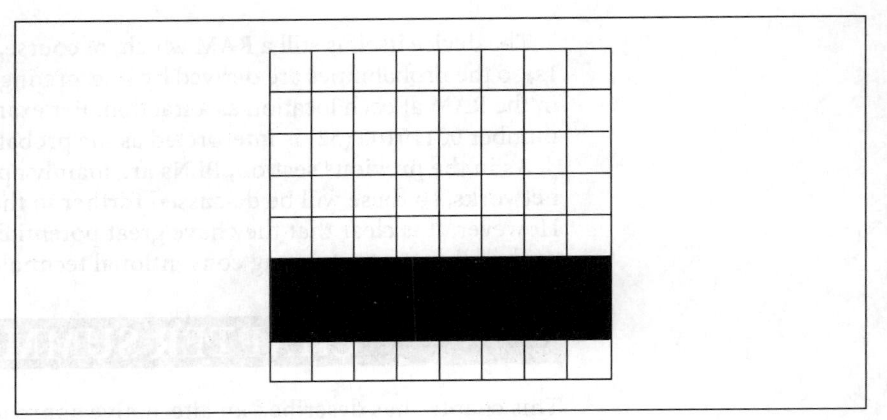

Figure 4.10 Test image for Question 3

2 Zero is represented by all 0s, and then as the number to be represented gets bigger, so one more digit has to become 1. To represent the figure 1 you need one 1, for 2 you need two 1s, for 3 you need three 1s and so on until for n you need n 1s. Therefore, with an n-bit code you can represent the numbers from 0 to n which means you have $n + 1$ different codes.

Table 4.4 Result of showing the WISARD the new horizontal bar (see answer to Question 3)

RAM	Address	V	H
1	0001	1	1
2	0001	1	1
3	0011	0	0
4	0011	0	0
5	0011	0	0
6	0000	1	1
7	0001	1	0
8	0001	0	1
9	0001	0	0
10	0000	1	1
11	0011	0	0
12	0001	1	1
13	0001	0	1
14	0111	0	1
15	0111	0	0
16	0111	0	1
	Total	6	9

3 Table 4.4 shows the addresses in the RAMs where 1s are stored, all other addresses containing 0. When the new image is shown to the WISARD, the addresses can be determined by comparing the image with the address map shown in Figure 4.6(b). The resulting addresses and the contents are shown in the following table. The total output for the vertical discriminator is 6, and the horizontal discriminator is 9. The WISARD would therefore correctly conclude that the bar is horizontal.

CHAPTER 5

Associative memory and feedback networks

CHAPTER OVERVIEW

This chapter shows how feedback can be used around a neural network to produce an associative memory. It explains:

- how to set the weights in a Hopfield network
- how to calculate the next states in a Hopfield network
- how to calculate the energy in a Hopfield network
- how to calculate the capacity of a Hopfield network
- how the Hamming network operates
- how the bidirectional associative memory (BAM) works

5.1 Associative memory

First, a review of what was said in Chapter 1. It was said that neural networks are basically pattern classifiers, and the previous chapters showed how this is achieved. There is another aspect of neural networks that needs including, however, as a kind of consequence of its pattern-classifying ability. This other property is its ability to associate one pattern with another.

Take the numeral classifier again. We have seen how it is possible to build a network that can be trained to classify the numerals 0 to 9. The network can then be shown variable or noisy examples of the numerals which it still correctly classifies. The output of the system that was discussed in Chapter 1 consisted of 10 lines connected to lamps which would turn on to indicate the class.

Now, it is not too difficult to imagine a situation where those 10 lines are used as the address input to another memory where good examples of the 10 numerals are stored. Then, when a hand-drawn number 3 arrives at the input, the output is a good quality version of the number 3, and similarly for all the other numerals.

What would have been constructed is an associative memory, where a number of different input patterns (all possible versions of the number 3 say) are associated with a particular output pattern (the good version of the number 3).

Two points need to brought out of this. The first is that the patterns that we want associated could be anything. For example, the input could be a numeral as before, and the output could be a spoken version of the number in any language.

Secondly, and probably more importantly, there does not have to be a separate memory where the output patterns are stored. During supervised training, the input patterns are presented to the system with the required output pattern. If training is successful, input patterns will produce the correct output patterns. So the system is acting as a pattern classifier and an associative memory at the same time.

One way of thinking about a neural network is therefore as a kind of memory where the information is stored not in its direct form, as in a RAM, but distributed over the network. The input patterns can be thought of as addresses which produce the desired outputs.

The ideal characteristics of an associative memory can be listed as follows (Pao, 1989):

- it should be able to store many associated pattern pairs
- it should accomplish this storage through a self-organising process
- it should store this information in a distributed manner
- it should generate the appropriate output response on receipt of the associated input pattern
- it should generate the correct output response even if the input pattern is distorted or incomplete
- it should be able to add new associations to the existing memory

Memories where the stored output patterns are different from the corresponding input patterns are called hetero-associative. A special case exists where the input and the output patterns are the same, in which case the memory is said to be auto-associative. At first sight this might seem to be a useless device, since it produces an output pattern when the same input pattern is presented. Its advantage is that it can also produce the correct output pattern when the input pattern is not perfect, for example, by being noisy or incomplete.

So these are the important properties of associative memories, but why do we need yet another type of neural network when the multi-layered perceptron and the Boolean networks are popular and have been shown to work? Two reasons are the biological implausibility of the networks described so far, and the time required to train the networks. The first reason is often used by researchers who believe that to achieve truly intelligent machines, anything that is not biologically plausible must be rejected. Since there is no evidence of anything like back-propagation or electronic memories in nature, these must be rejected. The second argument is perhaps more pragmatic, concerning itself with the problem of speed. Although back-propagation appears to work, it often takes a very long time to find to a valid set of weights. (This argument does not apply to Boolean networks.)

The solutions proposed in this chapter are feedback networks, where simple Hebbian learning is used, and training is achieved with a single calculation. Thus, they can be trained very quickly, and in some cases, they are biologically plausible. Inevitably, by solving one set of problems, a new set is introduced. These problems will also be described in this chapter. In the following chapter, some of the attempts at overcoming the new problems will be discussed.

5.2 The learning matrix

There are many ways of constructing associative memories. The first that will be described is an early example of an associative memory, called the learning matrix.

The learning matrix (Steinbuch, 1961; Steinbuch and Piske, 1963) is an electronic device that adapts the weighted connections between lines using a Hebbian learning rule. Figure 5.1 represents the matrix, with inputs, x_i, and outputs y_j.

The connections between the input and output lines are weighted, w_{ij}. During training, a pattern is presented at the input lines, and the required output pattern is placed on the output lines. The weights are incremented by a small amount if both the input line, x_i, and output line, y_j, are 1. Originally it was suggested that the weights should also be decremented if the input is 1 when the output is 0, but this was later dropped due to

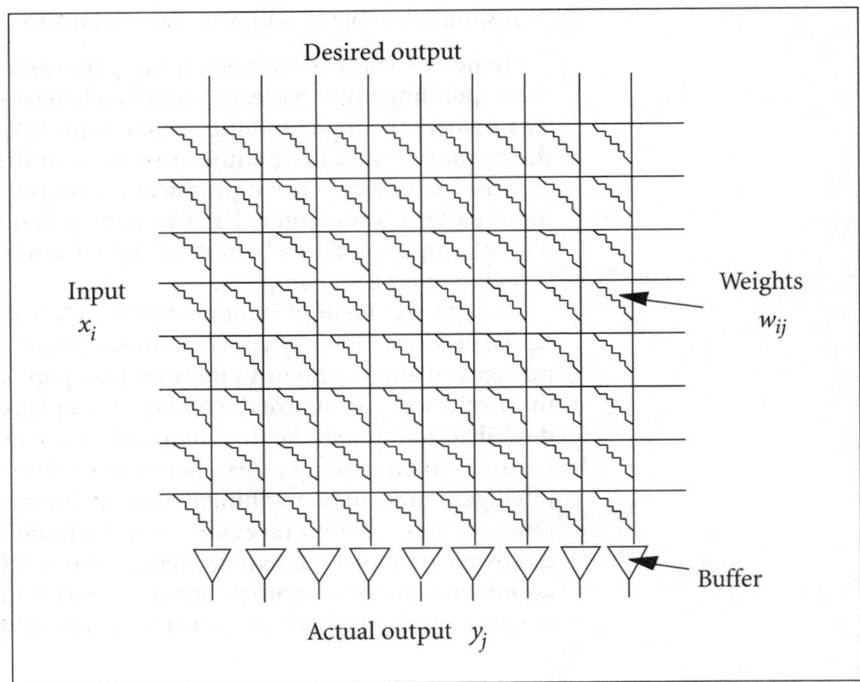

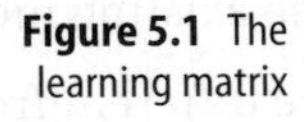
Figure 5.1 The learning matrix

technical difficulties. The matrix therefore uses the form of Hebbian learning that was described in Chapter 2.

If the weights are the conductances of the connections, and the inputs and outputs are voltages, then the current flowing from an input through a resistor is found using Ohm's law:

$$\text{current} = w_{ij}x_i$$

The total current flowing to the buffer at output, y_j, is:

$$\text{total current} = \sum_{i=0}^{n} w_{ij}x_i$$

So, the output voltage, y_j, from each buffer (unity gain amplifier) is:

$$y_j = z\sum_{i=0}^{n} w_{ij}x_i$$

where z is the output impedance of the buffer.

Clearly then, we have a similar equation to the weighted sum of the perceptron and ADALINE networks, including the term w_{0j} which provides an offset. The difference is in the training, which does not compare the desired and actual output but just uses the value of the desired output.

If there are P training pairs, then

$$w_{ij} = \eta\sum_{p=1}^{P} x_{ip}y_{jp}$$

where η is a small amount.

The learning matrix has been introduced because it is a physical realisation of the matrix notation that will be used. The operation of the learning matrix can be described mathematically as:

$$[\mathrm{Y}] = [\mathrm{X}][\mathrm{W}]$$

$$[\mathrm{W}] = \eta[\mathrm{X}]^{\mathrm{t}}[\mathrm{Y}]$$

where [X], [Y] and [W] are matrices, and $[\mathrm{X}]^{\mathrm{t}}$ is the transpose, as in Chapter 2.

In matrix notation, the input vector, [X], has P rows, labelled X1 to XP. These vectors represent the input patterns, so X1 is the first input pattern and XP is the last. Similarly, the associated output pattern vectors, [Y], has P rows, summarised as Y1 to YP. Notice that the term vectors are used to describe the inputs and outputs because matrix notation is being used. We can always convert the vectors back to a pattern and vice versa. In general, the set of input patterns, [X], will have P input vectors, each of which can be expanded into n values, x_{ip}. Therefore, x_{ip} is the ith input with the value from the pth pattern.

Let us do a simple example, where there are four inputs and four outputs, and the system is trained by showing it two input patterns and their corresponding output patterns. So, $n = 4$ and $P = 2$.

$$[X]=\begin{bmatrix}X1\\X2\end{bmatrix}=\begin{bmatrix}0&0&1&1\\0&1&0&1\end{bmatrix}\quad [Y]=\begin{bmatrix}Y1\\Y2\end{bmatrix}=\begin{bmatrix}0&0&0&1\\1&1&0&0\end{bmatrix}$$

This means that when 0101 is presented to the input, the desired output is 1100 and so on. So there are four inputs and four outputs.

$$w_{ij}=\eta\sum x_i y_j$$

$$[W]=\eta[X]^t[Y]$$

Let us make $\eta = 1$ for simplicity.

$$[X]^t=\begin{bmatrix}0&0\\0&1\\1&0\\1&1\end{bmatrix}\quad \text{so } [W]=\begin{bmatrix}0&0\\0&1\\1&0\\1&1\end{bmatrix}\begin{bmatrix}0&0&0&1\\1&1&0&0\end{bmatrix}$$

$$W=\begin{bmatrix}0&0&0&0\\1&1&0&0\\0&0&0&1\\1&1&0&1\end{bmatrix}$$

The matrix can be thought of as a pattern classifier that will give the correct class [Y] when an input pattern [X] appears, or as a pattern associator, associating [X] and [Y], or as a distributed memory which has stored the two patterns [Y] which are addressed by [X].

Now, when an input vector is multiplied by the matrix, [W], the correct output vector should appear, but it does not.

$$\begin{bmatrix}0&0&1&1\\0&1&0&1\end{bmatrix}\begin{bmatrix}0&0&0&0\\1&1&0&0\\0&0&0&1\\1&1&0&1\end{bmatrix}=\begin{bmatrix}1&1&0&2\\2&2&0&1\end{bmatrix}$$

The correct pattern is obtained by thresholding the output with a suitable positive value.

The matrix [W] is called the correlation matrix, or sometimes the association matrix or connection matrix. For a relatively small number of examples it has the ability to store the input-to-output relationship. In addition, it can give the correct output pattern even when some of the input pattern is corrupted. For example, if the input pattern 0011 is corrupted by replacing the first bit by 1, then:

$$[1 \quad 0 \quad 1 \quad 1]\begin{bmatrix} 0 & 0 & 0 & 0 \\ 1 & 1 & 0 & 0 \\ 0 & 0 & 0 & 1 \\ 1 & 1 & 0 & 1 \end{bmatrix} = [1 \quad 1 \quad 0 \quad 2]$$

The matrix is still capable of producing the correct output. Of course, this example was chosen well, and this is not foolproof, but it does demonstrate some of its abilities.

5.3 The Hopfield network

The learning matrix has some interesting properties:

- being able to store a number of input/output associations in one matrix
- being able to generalise to some extent, by recognising similar but not identical input patterns and producing the same output pattern
- following from the previous property, it has the ability to produce the correct output from incomplete or corrupted input patterns

However, it suffers from the same drawbacks as single-layer networks, that it cannot represent non-linearly separable functions. Also, it is restricted by the use of 0,1 values when, as seen earlier, there is an advantage to using –1,+1 because the weights can be incremented or decremented.

If the correlation matrix from the previous example is calculated again using –1,+1, the result looks like this.

$$[X'] = \begin{bmatrix} -1 & -1 & +1 & +1 \\ -1 & +1 & -1 & +1 \end{bmatrix} \quad [Y'] = \begin{bmatrix} -1 & -1 & -1 & +1 \\ +1 & +1 & -1 & -1 \end{bmatrix}$$

$$[X']^t = \begin{bmatrix} -1 & -1 \\ -1 & +1 \\ +1 & -1 \\ +1 & +1 \end{bmatrix} \text{ so } [W']^t = \begin{bmatrix} -1 & -1 \\ -1 & +1 \\ +1 & -1 \\ +1 & +1 \end{bmatrix} \begin{bmatrix} -1 & -1 & -1 & +1 \\ +1 & +1 & -1 & -1 \end{bmatrix}$$

$$W' = \begin{bmatrix} 0 & 0 & 2 & 0 \\ 2 & 2 & 0 & -2 \\ -2 & -2 & 0 & 2 \\ 0 & 0 & -2 & 0 \end{bmatrix}$$

$$\begin{bmatrix} -1 & -1 & +1 & +1 \\ -1 & +1 & -1 & +1 \end{bmatrix} \begin{bmatrix} 0 & 0 & 2 & 0 \\ 2 & 2 & 0 & -2 \\ -2 & -2 & 0 & 2 \\ 0 & 0 & -2 & 0 \end{bmatrix} = \begin{bmatrix} -4 & -4 & -4 & +4 \\ +4 & +4 & -4 & -4 \end{bmatrix}$$

A hard-limiter on the output would therefore produce the required output.

The Hopfield network (Hopfield, 1982) is a radical development of the learning matrix: it assumes that the outputs, [Y], are connected to the inputs, [X], as shown in Figure 5.2.

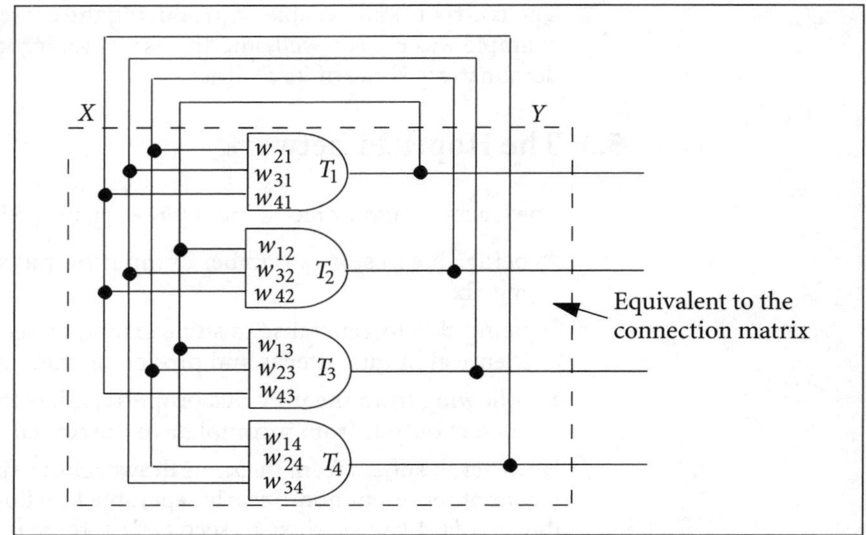

Figure 5.2 The Hopfield network

Thus feedback has been introduced into the network. The present output pattern is no longer solely dependent on the present inputs, but is also dependent on the previous outputs. Therefore the network can be said to have some sort of memory, as its outputs are some function of the current inputs and the previous outputs. Notice that, compared with the feedforward networks that have been described in earlier chapters, the Hopfield network has only one layer of neurons.

When presented with an input pattern, the final or desired output is not immediately produced but, instead, the output keeps changing as the network passes through several states until it converges to a stable state. In this description, 'stable' means that no further changes take place. Viewed in this light, one way of describing the function of the network is to say that the initial input pattern determines the initial state, and the desired output pattern determines the final state of the network. One crucial feature of the Hopfield network is that the outputs from each of the neurons must change asynchronously, that is, one at a time. The reason for this will be explained later when the 'energy' in the system will be discussed.

In the Hopfield network, inputs of 0 or 1 are usually used, but the weights are initially calculated after converting the inputs to –1 or +1 respectively.

The response of an individual neuron in the network is given by:

$$y_j = 1 \text{ if } \sum_{i=1, i \neq j}^{n} w_{ij} x_i > T_j$$

$$y_j = 0 \text{ if } \sum_{i=1, i \neq j}^{n} w_{ij} x_i < T_j$$

This means that for the jth neuron, the inputs from all other neurons are weighted and summed. Note that $i \neq j$, which means that the output of each neuron is connected to the input of every other neuron, but not to itself. The output is a hard-limiter which gives a 1 output if the weighted sum is greater than T_j and an output of 0 if the weighted sum is less than T_j. What happens if the weighted sum equals T_j? This is not clear in Hopfield's paper (Hopfield, 1982), so it will be assumed that the output does not change when the weighted sum is equal to T_j.

The weights are calculated in advance as a matrix, [W]. This means that there is still a learning phase, but, compared with feedforward networks, this is very short, since it involves just one calculation. When the weights are set like this, the network is effectively 'memorizing' the input patterns, where the patterns are stored as stable states.

Since the Hopfield network is auto-associative, the outputs are the same as the inputs, so the formula for calculating the weights is:

$$[\mathrm{W}] = [\mathrm{X}]^{\mathrm{t}}[\mathrm{X}] - [P]$$

where [X] is the matrix of input patterns coded as –1,+1 and $[P]$ is the unit matrix multiplied by P, the number of patterns in the training set. The unit matrix is introduced to ensure that there is no connection between an output of a neuron and one of its own inputs.

For example, if the patterns X1 = [0 0 1 1] and X2 = [0 1 0 1] are to be stored, first convert them to [–1–1+1+1] and [–1+1–1+1], then multiply each vector by itself and subtract the unit matrix multiplied by the number of patterns being stored, which in this case is 2.

$$\begin{bmatrix} -1 & -1 \\ -1 & +1 \\ +1 & -1 \\ +1 & +1 \end{bmatrix} \begin{bmatrix} -1 & -1 & +1 & +1 \\ -1 & +1 & -1 & +1 \end{bmatrix} - \begin{bmatrix} 2 & 0 & 0 & 0 \\ 0 & 2 & 0 & 0 \\ 0 & 0 & 2 & 0 \\ 0 & 0 & 0 & 2 \end{bmatrix} = \begin{bmatrix} 0 & 0 & 0 & -2 \\ 0 & 0 & -2 & 0 \\ 0 & -2 & 0 & 0 \\ -2 & 0 & 0 & 0 \end{bmatrix}$$

This is the starting point for a Hopfield network. We have:

$$w_{11} = 0 \quad w_{12} = 0 \quad w_{13} = 0 \quad w_{14} = -2$$

$$w_{21} = 0 \quad w_{22} = 0 \quad w_{23} = -2 \quad w_{24} = 0$$

$$w_{31} = 0 \quad w_{32} = -2 \quad w_{33} = 0 \quad w_{34} = 0$$

$$w_{41} = -2 \quad w_{42} = 0 \quad w_{43} = 0 \quad w_{44} = 0$$

The symmetry can be seen, for example, where $w_{14} = w_{41} = -2$, which means that the connection between neuron 1 and neuron 4 has a weight of −2 associated with it.

Thresholds also need to be calculated. This could be included in the matrix by assuming that there is an additional neuron, called neuron 0, which is permanently stuck at 1. All other neurons have input connections to this neuron's output with weights, w_{01}, w_{02} etc. This provides an offset which is added to the weighted sum. The relationship between the offset and the threshold T_j is therefore:

$$T_j = -w_{0j}$$

The value of w_{0j} is calculated from the original input patterns. As usual, the weights are found using the formula $[W] = [X]^t[Y]$, but in this instance, the output [Y] is just the output of neuron 0 which is permanently stuck at 1, so the formula becomes:

$$[W_0] = [X]^t[Y_0]$$

In the current example, the two patterns that are being learnt by the network are [−1−1+1+1] and [−1+1−1+1], and y_0 is permanently stuck at +1, so the offsets are calculated as follows:

$$[W_0] = \begin{bmatrix} -1 & -1 \\ -1 & +1 \\ +1 & -1 \\ +1 & +1 \end{bmatrix} \begin{bmatrix} +1 \\ +1 \end{bmatrix} = \begin{bmatrix} -2 \\ 0 \\ 0 \\ +2 \end{bmatrix}$$

Alternatively, these weights could be converted to thresholds to give:

$$T_1 = 2,\ T_2 = 0,\ T_3 = 0,\ T_4 = -2$$

If the network is shown either of the two original patterns it will produce the correct response. This means that the patterns will appear as stable outputs. The way that this is done is to connect the inputs to the network, so that they force the output lines to have the same values as the input. This defines the initial state. The inputs are removed, and these output values are fed back one at a time to the inputs and new outputs calculated one at a time. This is repeated until the outputs stop changing, in which case the network has converged to its final state.

The example that has been shown would be correct immediately, which means that simply presenting the input pattern would immediately put the network into a stable state so that no further change takes place. Figure 5.3 shows the case where the input pattern is 0011.

The Hopfield network's ability really comes to light when it is shown a corrupted pattern. Take, for example, the pattern 0010, shown in Figure 5.4.

Again, the input pattern is imposed on the output. Nothing has been said about the order in which the outputs are updated. It is generally assumed that this is random, but over a period of time the neurons will update the same number of times. However, the order can make a

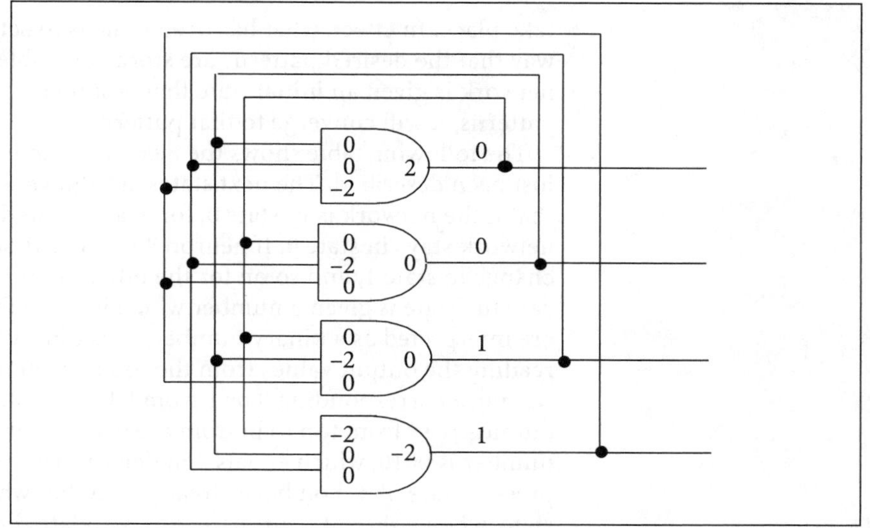

Figure 5.3 Hopfield network with a stable output of 0011

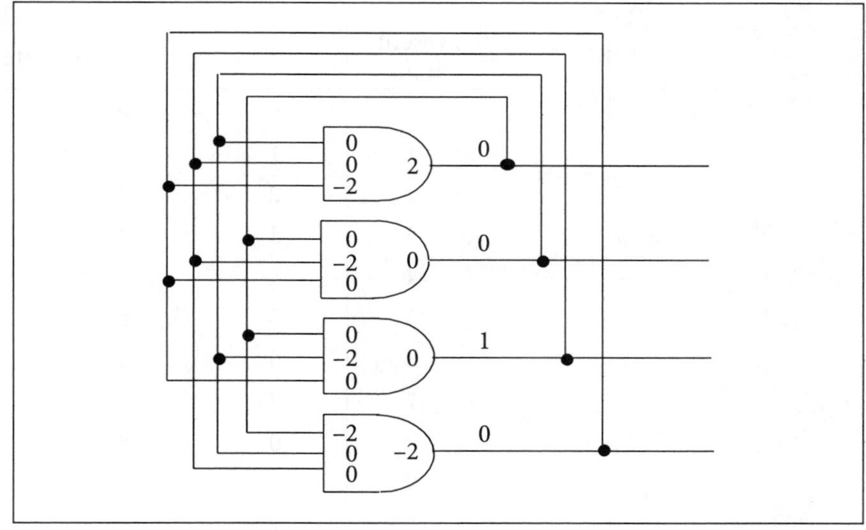

Figure 5.4 Hopfield network with an output of 0010

difference. First update the outputs starting at neuron 1 at the top. The output values are fed back so that the sum produced in neuron 1 is 0. The sum is less than the threshold so it stays at 0.

Next, neuron 2 updates which also does not produce any changes, and similarly for neuron 3. However, when neuron 4 updates, the weighted sum is 0, which causes the output to change to 1 since it is greater than the threshold in this neuron. The state becomes 0011. No more changes take place, and the output has converged to the state which was one of the original patterns. Furthermore, it is the pattern which is 'closest' to the corrupted pattern.

This illustrates the way in which the network progresses through a series of states until one is reached which is stable and no further changes take place. In effect, what has been done is to set the weights in such a way that the desired patterns are stored as stable states. When the network is given an initial state that is similar to one of the stored patterns, it will converge to that pattern.

The following table shows the next state table for the example that has just been described. The next states are shown with a particular neuron so that if the network is in state 0, for example, and neuron 1 updates, the network stays in state 0. If neuron 4 updates first, then state 0 would change to state 1, and so on for the other neurons and next states. In each case the state is given a number which is value that you get if the outputs are interpreted as a binary number. This binary number is found by reading the output values from the top neuron to the bottom neuron, and writing a corresponding 0 or 1 from left to right. For example, if the outputs read from top to bottom are 0, 0, 1, 0, the corresponding binary number is 0010, which equals 2 in denary. The neuron therefore has a present state of 2. You have already seen that when the neuron is in this state, when either of neurons 1, 2 or 3 update there is no change. However, when neuron 4 updates, the state changes to 3.

Present state	Next state			
0	0	0	0	1
1	1	1	1	1
2	2	2	2	3
3	3	3	3	3
4	4	4	4	5
5	5	5	5	5
6	6	2	4	7
7	7	3	5	7
8	0	8	8	8
9	1	9	9	9
10	2	10	10	10
11	3	11	11	11
12	4	12	12	12
13	5	13	13	13
14	6	10	12	14
15	7	11	13	15
neuron	1	2	3	4

It can be seen from the figure and the table that states 1, 3 and 5 do not change, so that after entering one of those states the network stays there.

It can also be seen that no matter what state the network starts from, it will end up in one of the three stable states. The aim of the network was to store only two stable states, namely 3 and 5. The other stable state is not wanted and can produce incorrect or undesirable outputs.

One way of avoiding undesirable stable states is to increase the number of neurons. The number of neurons that are needed before errors in recall become severe can be quite large. The figure quoted by Hopfield is that for n neurons only about $P = 0.15n$ patterns can be stored (Hopfield, 1982). The implication of this is that to store just two patterns, at least 14 neurons are required, so problems with 14 inputs or fewer are not worth trying.

Other attempts at quantifying the number of patterns that can be stored have produced the following equations:

$$P = 0.138n \qquad \text{(Crisanti } et\ al.\text{, 1986)}$$

$$P = \frac{n}{2\ln n} \qquad \text{(McEliece } et\ al\text{, 1987)}$$

$$P = \frac{2}{2\ln n + \ln\ln n} \qquad \text{(Amari and Maginu, 1988)}$$

Let us return to the example of the 10 numerals again. The patterns consist of 64 binary pixels, so we need 64 neurons, and there are 10 classes. Using the above formula, $P = 0.15n$, we need 10/0.15 neurons to store 10 patterns, which turns out to be 67 neurons. Using 64 is close, but not quite enough, so we should expect to get some errors when this problem is implemented. Errors mean that on some occasions the stable output that is obtained is not the correct one, and so inputs are misclassified.

The other three estimates show that for $P = 10$, $n = 73$, 90 or 110 neurons are needed for perfect recall. This would suggest that the problem is incapable of being solved with just 64 neurons.

Another major problem is in the choice of patterns to be stored. A purely arbitrary set of patterns is unlikely to produce good results. Since the aim is to store 'good' examples which should be recalled when the system is presented with 'bad' examples, the choice of stored patterns should be made to ensure good operation. Furthermore, if extra bits are added to patterns to increase the value of n then these values have to be carefully chosen. In general the values should be chosen in such a way that the difference between the patterns is increased.

It has been shown (Grant and Sage, 1986) that poor performance in the Hopfield network is associated with large cross-correlation coefficients between the ideal patterns that are stored. Cross-correlation coefficients are a measure of similarity between patterns. It is suggested that instead of constructing the connection matrix using the original set of patterns, a new set of vectors should be used, derived from the ideal vectors, but which are orthogonal. This is supported in situations where n-bit orthogonal codes are stored (Selviah *et al.*, 1989), when it can be shown that a maximum of n different codes can be stored. This agrees with the figure that has been reported elsewhere (Abu-Mostafa and St Jaques, 1985).

5.4 Energy

Hopfield, among others (Little, 1988), has made an analogy between training the neural network and minimising a value which has been interpreted as its global 'energy'. To understand this we need to define what is meant by the energy of the network.

Imagine a neuron with a particular input that has a large positive weight associated with it. Then, when that input is firing (equal to 1) it is likely that the neuron will also fire. So the large positive weight means that the neuron fires easily, only requiring a stimulus on one of its inputs. Alternatively, an input with a small positive weight will probably not be able to make the neuron fire on its own. Instead, several inputs with small positive weights are probably required to overcome the offset and cause the neuron to fire. So in this case, a lot of stimulus or 'energy' is required. Finally, inputs with negative weights will inhibit the neuron, so even more energy is required on the other inputs to cause the neuron to fire.

From this discussion, it can be seen that the 'energy' in a neural network is related to the negative sum of all the weighted inputs of the neurons. For a single neuron, the energy at a particular time is defined as:

$$E = -y\sum_{i=0}^{n} w_i x_i$$

This has been stated as an equality, without any other constants, because it does not effect the arguments that is going to be used from now on. It can be justified by saying that the quantity that has been called E is only analogous to energy and is not intended as an accurate quantitative measure.

The variables x and y are binary and can have the values 0,1. The change in energy when the output, y, changes, is ΔE. This can be calculated using the following equation:

$$\Delta E = -\Delta y\sum_{i=0}^{n} w_i x_i$$

If the aim is to minimise E, then one way that this could be achieved is to make sure that ΔE is always negative. First, if the output changes from 0 to 1, Δy is positive and the change in energy is:

$$\Delta E = -\sum_{i=0}^{n} w_i x_i$$

Secondly, if the output changes from 1 to 0, Δy is negative and the change in energy is:

$$\Delta E = \sum_{i=0}^{n} w_i x_i$$

This value can be made to be always negative if we ensure that the output is 1 when $\Sigma w_i x_i$ is greater than 0 and the output is 0 when $\Sigma w_i x_i$ is less than 0. This, of course, is precisely what a neuron does, provided that a set of weights exists that satisfies these equations.

A neuron starts off with an arbitrary set of weights. Then, the weights are adjusted so that the appropriate function is represented by the neuron. Adjusting the weights to produce the correct function is equivalent to minimising the energy of the neuron, the minimum occurring when the function is correctly stored.

For networks of neurons, the total energy is the sum of all the energies of the individual neurons. For an individual neuron in that network, the procedure for minimising the energy is the same as the argument that has just been put forward for an isolated neuron, provided that its weights are adjusted while all other neurons are held constant.

It is important that each neuron updates while all others stay fixed, so that the inputs x_i, which are taken from the outputs of all the other neurons, are all constant. This is why it was said earlier that the changes were asynchronous.

In the Hopfield network, the weights are calculated using the matrix method described earlier. After the weights have been established, an input pattern is imposed on the output, and their values get fed back to the inputs of each of the neurons. At this point, it is likely that many of the neurons will produce a weighted sum which would force the output to change its value. So, for example, a particular neuron might have an output of 1, and when the weighted sum is calculated, it is found to be less than 0. Thus, when it is this particular neuron's turn to update, the output will change from 1 to 0. Just prior to changing, the energy in that neuron with $y = 1$ can be calculated as follows:

$$E = -y \sum_{i=0}^{n} w_i x_i = -\sum_{i=0}^{n} w_i x_i > 0$$

After the neuron has updated, y will be 0 and the energy will be:

$$E = -y \sum_{i=0}^{n} w_i x_i = 0$$

The energy has therefore dropped by the value of the weighted sum.

One of the problems of the Hopfield network is that it has a tendency to find local energy minima rather than the global minimum. Some of these local minima could correspond to the unwanted stable states mentioned earlier and therefore the outputs are incorrect. It has been suggested (Rumelhart and McClelland, 1988) that the cause for this may be due to the fact that the Hopfield network uses hard-limiters. The neurons in the network are therefore forced into making binary decisions before they have a chance to communicate with each other. Hopfield has proposed a continuous version of his network (Hopfield, 1984) which, he claims, has overcome the problem of settling in local minima.

5.4.1 Ising spin-glass

A number of authors have demonstrated the similarity between a neuron, which can have an output of –1 or +1, and the Ising model of ferromagnetism, in which the unpaired electron spin can also have a value of –1 or +1, particularly when applied to magnetic alloys or spin-glasses as they are known (Sherrington, 1989). The benefit of making this analogy is that there is a wealth of understanding and analysis available on the subject of statistical mechanics (from which the Ising model is taken) which can be applied to some types of neural network.

In essence, suppose there are a chain of points with coordinates *i* and that at each point is a particle. A particle can be in either of two states, called its spin, and signified by a number, *s*, which can have the value of +1 or –1 corresponding to spin-up or spin-down respectively. The energy of the system can be written as:

$$E = -\frac{J}{2}\sum C_{ij} s_i s_j$$

where C_{ij} is the connection matrix which has elements with a value of 1 if the two particles are neighbours and 0 otherwise.

This was the model proposed by Ising for a substance exhibiting ferromagnetic properties (Ising, 1925). J is a positive constant which is equal to the contribution to the total energy given by a particular pair of particles. If the spins are aligned, the energy decreases and, in the limit, when all particles are aligned, the energy is at its minimum. The model can be made much more complex with the inclusion of the effects of external magnetic fields on the spin of the electrons.

There is not enough space in this book to develop this in any depth so this topic will be left as further reading for anyone who is interested. In particular the *Proceedings of the Heidelberg Colloquium on Spin Glasses, Optimisation and Neural Networks* (van Hemmen and Morgenstern, 1986) covers many aspects of the topic.

5.5 The Hamming network

A variation on the Hopfield network is the Hamming network shown in Figure 5.5 (Lippmann, 1987). This is a two-layer network, where the first layer is used to calculate a score, and the second layer is used to select the maximum. The second layer is essentially a Hopfield network with the exception that feedback is allowed from each neuron to itself.

Each neuron in the first layer is set up to give the maximum response to one of the training patterns. It does this by setting the weights in the *i*th neuron equal to half of the value of the *i*th input pattern. The weights of the *i*th neuron in the first layer are therefore set to:

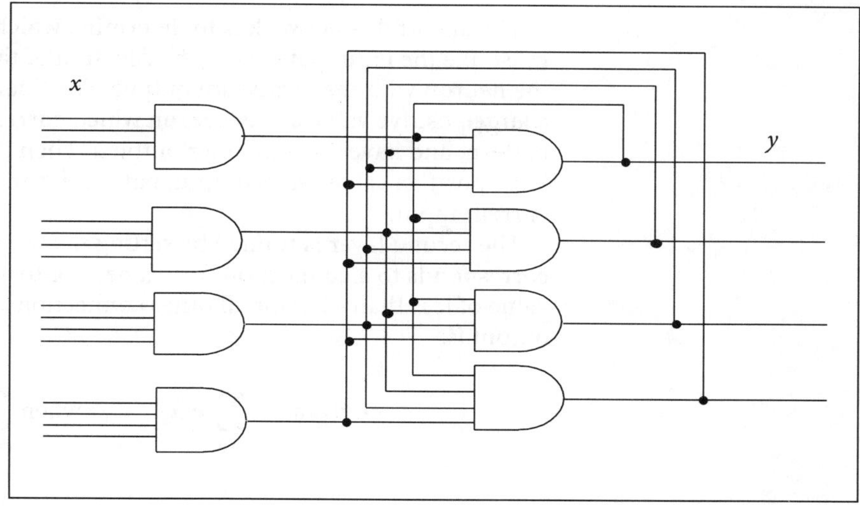

Figure 5.5 A Hamming network

$$w_{ij} = \frac{x_{ij}}{2}$$

where x_{ij} is the jth bit of the ith pattern, having a value of –1 or +1. The offset, w_{i0} is set to $n/2$, where n is the number of bits in the pattern.

For example, suppose the ith pattern is –1 –1 +1 +1. The weights of neuron i will be:

w_{i0}	w_{i1}	w_{i2}	w_{i3}	w_{i4}
+2.0	–0.5	–0.5	+0.5	+0.5

The weighted sum therefore equals 4. This value appears at the output since there are no non-linear output functions in these neurons.

Each of the neurons in the first layer gives a maximum response of n to one particular pattern, and a smaller amount to other patterns. So each neuron can be said to represent one of the training set of patterns. If there are P patterns, there will be P neurons in both layers.

When a new pattern is shown to the first layer, the output from each of the neurons in the first layer will be a measure of the Hamming distance from the patterns in the training set. The Hamming distance between two patterns is a measure of the number of bits that are different between the two patterns. For example, if an input pattern of –1 +1 +1 +1 is presented to the input, the ith neuron that was seen above will give an output of:

$$2 + (-1 \times -0.5) + (+1 \times -0.5) + (+1 \times 0.5) + (+1 \times 0.5) = 3$$

which is 1 less than the maximum of 4. This is because the pattern has 1 bit different, which is a Hamming distance of one. Note that if the input pattern is the inverse of one of the stored patterns, the neuron that corresponds to that pattern will produce an output of 0. This means that the outputs of all the neurons will range from 0 to n, and never be negative.

The aim of this network is to determine which of the stored patterns is most like the input pattern, so the function of the second layer is to select the neuron with the maximum output. The final output, y, should contain a large positive value at the neuron which corresponds to the nearest pattern, and have 0s at all other outputs. Then if neuron i ends up with a large positive value on its output, pattern i is the nearest pattern to the current input.

The second layer is trained by setting the weights to 1 if the weight corresponds to a connection from a neuron to itself, and a small negative value of less than $-1/P$ for all other connections. Offsets are all 0, and the output is:

$$\text{output} = \sum_{i=0}^{P} w_i x_i \qquad \text{when } \sum_{i=0}^{P} w_i x_i > 0$$

$$\text{output} = 0 \qquad \text{when } \sum_{i=0}^{P} w_i x_i < 0$$

During the operation of the network, the outputs are initially clamped to the value of the outputs of the first layer. Then the second layer is allowed to iterate. Initially the outputs of the second layer will be equal to the scores produced by the first layer. Then, because of the small negative weights between neurons, the outputs will gradually diminish. At some point, all of the outputs will have reached 0 except one – the output that had the largest value to start with. At this point the iterations stop because there will be no further changes. The remaining output corresponds to the neuron that contains the pattern that is nearest to the current input.

This is a rather strange network since it performs a function which would be far more easily implemented in a conventional algorithm. However, it demonstrates two properties of neural networks. The first is in the first layer where a neuron's ability to give a large output response to an incoming pattern is achieved by essentially storing a copy of that pattern as its weights. The second property is in the second layer, where the neurons interact to inhibit each other, with the result that a 'winner takes all' function is achieved. Both of these properties are used extensively in self-organising networks, which will be discussed in a later chapter.

5.6 Bidirectional associative memory (BAM)

A bidirectional associative memory (BAM) is very similar to a Hopfield network, but has two layers of neurons (Kosko, 1988), as shown in Figure 5.6. It uses the connection matrix described at the beginning of this chapter. The connection matrix was described as a means of calculating the outputs given some inputs. In the BAM, the connection matrix is also used to calculate the inputs given a set of outputs.

Figure 5.6 BAM

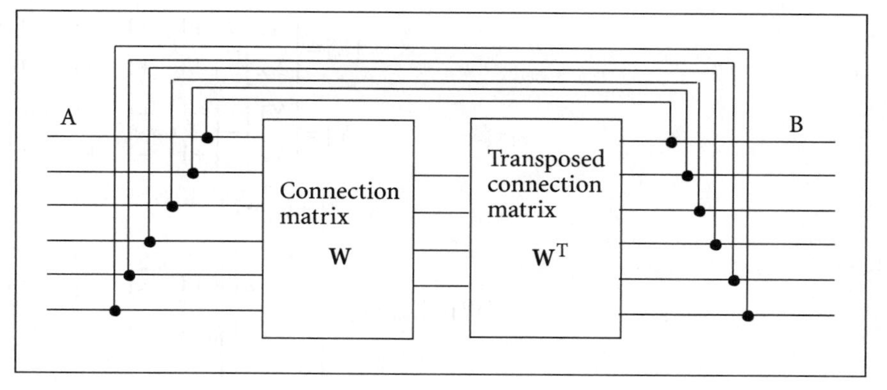

The connection matrix is constructed as before, by taking the product of the two input and associated output pattern matrices, [X] and [Y], which contain elements having the value of –1 or +1.

$$[W] = [X]^t[Y]$$

Unlike the Hopfield network, the diagonal of the connection matrix is left intact, so the unit matrix multiplied by *P*, [*P*], is not subtracted, and secondly the number of bits in the input pattern need not be the same as the output pattern, so the connection matrix is not necessarily square. The BAM operates by presenting an input pattern, [A], and passing it through the connection matrix to produce an output pattern, [B]. So:

$$[B(k)] = f([A(k)][W])$$

where k indicates time. The matrices $[A(k)]$ and $[B(k)]$ are equivalent to [X] and [Y] respectively, but with the values of 0 and 1 instead of –1 and +1. The outputs of the neurons are produced by the function $f(\ )$ which, like the Hopfield network, is a hard-limiter with a special case at 0. This output function is defined as follows:

$b(k+1) = 1$ if the corresponding element of $[A(k)][W] > 0$

$b(k+1) = 0$ if the corresponding element of $[A(k)][W] < 0$

$b(k+1) = b(k)$ if the corresponding element of $[A(k)][W] = 0$

The output, [B], is then passed back through the connection matrix to produce a new input pattern, [A].

$$[A(k+1)] = f([B(k)][W]^t)$$

The [A] and [B] patterns are passed back and forth through the connection matrix in the way just described until there are no further changes to the values of [A] and [B].

To illustrate this, here is an example (Kosko, 1987).

$$[A] = \begin{bmatrix} A1 \\ A2 \end{bmatrix} = \begin{bmatrix} 1 & 0 & 1 & 0 & 1 & 0 \\ 1 & 1 & 1 & 0 & 0 & 0 \end{bmatrix} \quad [B] = \begin{bmatrix} B1 \\ B2 \end{bmatrix} = \begin{bmatrix} 1 & 1 & 0 & 0 \\ 1 & 0 & 1 & 0 \end{bmatrix}$$

Converting to –1, +1 notation:

$$[X]=\begin{bmatrix} X1 \\ X2 \end{bmatrix}=\begin{bmatrix} +1 & -1 & +1 & -1 & +1 & -1 \\ +1 & +1 & +1 & -1 & -1 & -1 \end{bmatrix}$$

$$[Y]=\begin{bmatrix} Y1 \\ Y2 \end{bmatrix}=\begin{bmatrix} +1 & +1 & -1 & -1 \\ +1 & -1 & +1 & -1 \end{bmatrix}$$

$$[W]=[X]^t[Y]=\begin{bmatrix} +1 & +1 \\ -1 & +1 \\ +1 & +1 \\ -1 & -1 \\ +1 & -1 \\ -1 & -1 \end{bmatrix}\begin{bmatrix} +1 & +1 & -1 & -1 \\ +1 & -1 & +1 & -1 \end{bmatrix}=\begin{bmatrix} 2 & 0 & 0 & -2 \\ 0 & -2 & 2 & 0 \\ 2 & 0 & 0 & -2 \\ -2 & 0 & 0 & 2 \\ 0 & 2 & -2 & 0 \\ -2 & 0 & 0 & 2 \end{bmatrix}$$

This matrix, [W], now has each of the relationships between [X] and [Y] stored. To test this, the equation should be satisfied for each input.

$$[Bi] = f([Ai][W])$$

For example, [A1]:

$$[1 \quad 0 \quad 1 \quad 0 \quad 1 \quad 0]\begin{bmatrix} 2 & 0 & 0 & -2 \\ 0 & -2 & 2 & 0 \\ 2 & 0 & 0 & -2 \\ -2 & 0 & 0 & -2 \\ 0 & 2 & -2 & 0 \\ -2 & 0 & 0 & 2 \end{bmatrix}=[4 \quad 2 \quad -2 \quad -4]$$

After applying the non-linear function, $f(\ )$, the output is [1 1 0 0] which equals [B1].

Similarly, it should be possible to get [A*i*] from [B*i*] using the following equation:

$$[Ai] = f([Bi][W]^t)$$

For example, try [B2].

$$[1 \quad 0 \quad 1 \quad 0]\begin{bmatrix} 2 & 0 & 2 & -2 & 0 & -2 \\ 0 & -2 & 0 & 0 & 2 & 0 \\ 0 & 2 & 0 & 0 & -2 & 0 \\ -2 & 0 & -2 & 2 & 0 & 2 \end{bmatrix}=[2 \quad 2 \quad 2 \quad -2 \quad -2 \quad -2]$$

After applying the non-linear function, $f(\ )$, the output is [1 1 1 0 0 0], which equals [A2]. In this example, the system is immediately in a stable state. In other examples, the inputs are allowed to pass back and forth through the network until a stable state is reached, when the correct output pattern should be produced.

The BAM therefore uses the input pattern as the initial state of the system, just like other feedback networks such as the Hopfield network, and ends up in a stable state which is intended to be the desired output pattern. It achieves this by passing the data back and forth through the network, and therefore offers an alternative to the feedback mechanism. At each pass, gradient descent is again being applied in terms of the energy of the network. The change in either the input or the output pattern always produces a lower energy state.

CHAPTER SUMMARY

In this chapter the concept of an associative memory has been introduced. Central to these ideas is the connection matrix, which was used in the learning matrix, the Hopfield network, and the bidirectional associative memory (BAM). In the case of the Hopfield network, feedback is used around a network of perceptron-like neurons. Feedback creates a system which uses the input and output patterns to represent some of its states. Pattern classification occurs when an input pattern causes a sequence of changes ending in a stable state which corresponds to the desired output pattern or class.

Training in the Hopfield network was shown to be a gradient descent in terms of the energy of the system, with the input/output relationship being stored in energy wells or minima. This is also true of the BAM networks.

Associative memories are used in pattern classification, but are not as popular as the multi-layered perceptron. This is mainly due to the associative memory's ability to give incorrect outputs, caused by arriving at a state which is a local minimum in terms of the energy of the system. In the next chapter, the problem of local minima is discussed in more depth, and some attempts at overcoming this problem are described.

SELF-TEST QUESTIONS

1 (a) Using each of the formulae for the storage capacity of a Hopfield network, plot a graph of storage capacity, P, versus the number of neurons, n, for values of n up to 20.

(b) What is the minimum number of neurons that you would need if you wanted to reliably store three patterns?

(c) If the three patterns that you want to store are 8 bit patterns, how many neurons do you need to reliably store them?

2 Find the weights and thresholds for a Hopfield network that stores the patterns 001 and 011.

3 (a) For the network described in Figure 5.7, draw a state table showing what the next state will be when each of the neurons updates. Use the update rule that the output does not change if the weighted sum is equal to the threshold, and remember that the inputs and outputs have the values 0 and 1.

(b) Are states 1 and 3 stable?

4 For the network described in Figure 5.7, draw a state table showing the energy of all of the states.

5 A BAM is trained using the following input and output patterns:

Inputs	Outputs
000010010000010	01
000010000010000	10
000100100100000	11

Find the weights that would be generated for the BAM network, and check that the input patterns generate the corresponding output patterns.

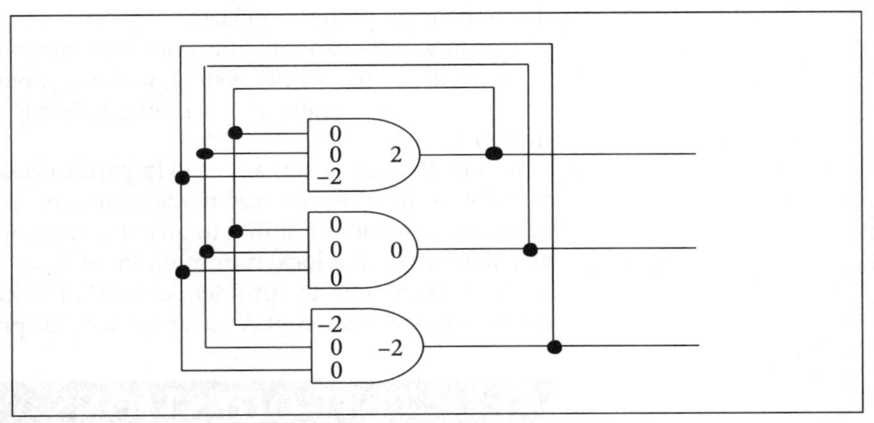

Figure 5.7 Hopfield network for Questions 3 and 4

SELF-TEST ANSWERS

1 (a) See Figure 5.8.

(b) 20, as indicated for $P = 3$ in the graph above.

(c) Since the patterns are 8 bit, you may expect to use 8 neurons. However, with $n = 8$ the network can only just reliably store one pattern. The previous answer suggested at least 20 neurons are required, so to store these three 8 bit patterns reliably, you would have to 'pad them out' to 20 bits, which means adding a redundant 12 bits to the data. The values chosen for the extra bits should attempt to make the patterns as different from one another as possible.

2 First convert the patterns to –1 and +1, so that they are –1–1+1 and –1+1+1. Next, arrange them as a matrix, and find the transpose of the matrix.

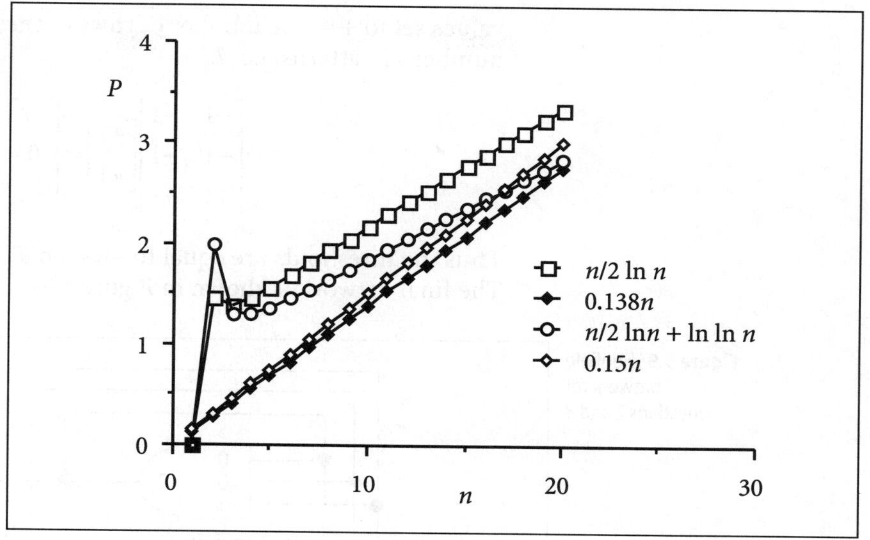

Figure 5.8 Capacity of a Hopfield network versus number of inputs

The matrix is $\begin{bmatrix} -1 & -1 & +1 \\ -1 & +1 & +1 \end{bmatrix}$ and the transpose is $\begin{bmatrix} -1 & -1 \\ -1 & +1 \\ +1 & +1 \end{bmatrix}$

Next, multiply the transpose by the original matrix, and subtract the unit matrix multiplied by the number of patterns, i.e. 2.

$$\begin{bmatrix} -1 & -1 \\ -1 & +1 \\ +1 & +1 \end{bmatrix}\begin{bmatrix} -1 & -1 & +1 \\ -1 & +1 & +1 \end{bmatrix} - \begin{bmatrix} 2 & 0 & 0 \\ 0 & 2 & 0 \\ 0 & 0 & 2 \end{bmatrix} = \begin{bmatrix} 0 & 0 & -2 \\ 0 & 0 & 0 \\ -2 & 0 & 0 \end{bmatrix}$$

The weights are therefore:

$$w_{11} = 0 \quad w_{12} = 0 \quad w_{13} = -2$$
$$w_{21} = 0 \quad w_{22} = 0 \quad w_{23} = 0$$
$$w_{31} = -2 \quad w_{32} = 0 \quad w_{33} = 0$$

Note that the connection from a neuron to itself has a weight of 0. This can either be shown as a weight of 0 in the Hopfield network, or the connection can be left out altogether.

The thresholds are calculated by multiplying the transposed matrix by a single column matrix of all +1s to get the values of w_{0i}. This is essentially a short cut. You could be more long-winded and start of by assuming that there is an additional neuron in the network which has a constant +1 output. You would then have to calculate the extra weight in each neuron, but would also have to calculate the weights in this additional neuron. Since this neuron has a constant output, its weights are irrelevant. So, to save effort, we just calculate the additional values of w_{0i} using a single column matrix with all the values set to +1. The number of rows of the matrix is equal to the number of patterns, i.e. 2.

$$\begin{bmatrix} -1 & -1 \\ -1 & +1 \\ +1 & +1 \end{bmatrix} \begin{bmatrix} +1 \\ +1 \end{bmatrix} = \begin{bmatrix} -2 \\ 0 \\ +2 \end{bmatrix}$$

Thus the thresholds are equal to $-w_{0i}$, so $T_1 = +2$, $T_2 = 0$, and $T_3 = -2$. The final network is shown in Figure 5.9.

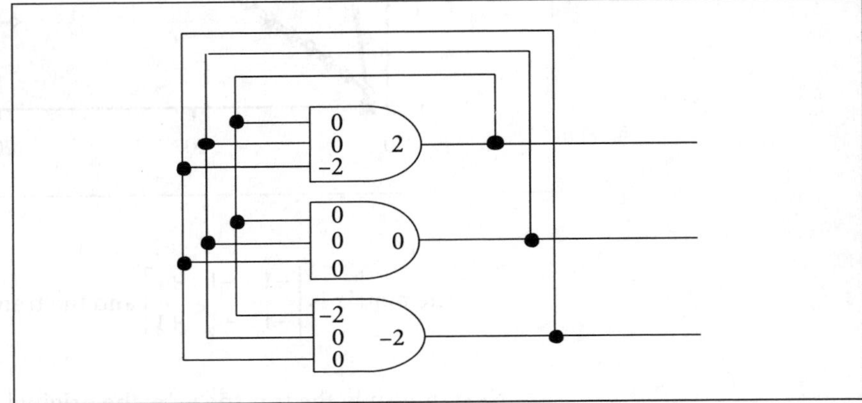

Figure 5.9 Hopfield network for Questions 2 and 3

3 (a) The following table shows all of the possible states that the network can be in, and all of the possible next states that it can change to.

state	next state		
0	0	0	1
1	1	1	1
2	2	2	3
3	3	3	3
4	0	4	4
5	1	5	5
6	2	6	6
7	3	7	7
	1	2	3
		neuron	

(b) Yes, both states 1 and 3 are stable. In effect, the network has learned to store both of the patterns. This is fortunate, as the storage capacity graph shown in Question 1 showed that with $n = 3$ the storage capacity is virtually zero.

4 The following table shows all of the possible states and their energy values. Note that states 1 and 3 have the lowest values of energy.

state	E
0	0
1	–2
2	0
3	–2
4	+2
5	+4
6	+2
7	+4

5 First convert the patterns to –1 and +1, and transpose the matrix of input patterns. Then multiply the two matrices to get the matrix of weights.

$$\begin{bmatrix} -1 & -1 & -1 \\ -1 & -1 & -1 \\ -1 & -1 & -1 \\ -1 & -1 & +1 \\ +1 & +1 & -1 \\ -1 & -1 & -1 \\ -1 & -1 & +1 \\ +1 & -1 & -1 \\ -1 & -1 & -1 \\ -1 & -1 & +1 \\ -1 & +1 & -1 \\ -1 & -1 & -1 \\ -1 & -1 & -1 \\ +1 & -1 & -1 \\ -1 & -1 & -1 \end{bmatrix} \begin{bmatrix} -1 & +1 \\ +1 & -1 \\ +1 & +1 \end{bmatrix} = \begin{bmatrix} -1 & +1 \\ -1 & +1 \\ -1 & +1 \\ +1 & +1 \\ -1 & -1 \\ -1 & -1 \\ +1 & +1 \\ +3 & +1 \\ -1 & -1 \\ +1 & +1 \\ +1 & -3 \\ -1 & -1 \\ -1 & -1 \\ -3 & +1 \\ -1 & -1 \end{bmatrix}$$

Next, try multiplying the input patterns with the weights to see if the resulting output patterns are correct. For example, the first pattern gives values of –7 and +1 which, when thresholded, gives an output pattern of 01, which is correct.

$$[0\ 0\ 0\ 0\ 1\ 0\ 0\ 1\ 0\ 0\ 0\ 0\ 0\ 1\ 0]\begin{bmatrix} -1 & +1 \\ -1 & +1 \\ -1 & +1 \\ +1 & +1 \\ -1 & -1 \\ -1 & -1 \\ +1 & +1 \\ +3 & +1 \\ -1 & -1 \\ +1 & +1 \\ +1 & -3 \\ -1 & -1 \\ -1 & -1 \\ -3 & +1 \\ -1 & -1 \end{bmatrix} = [-7\ \ +1]$$

The other two input patterns get values of 0 –4, which could become 00 or 10 depending on the value of the previous output, and 3 3 which becomes 11 after thresholding which is correct.

CHAPTER 6

Statistical neural networks

CHAPTER OVERVIEW

This chapter gives a description of feedback and feedforward networks that make use of statistics. It explains:

- how probabilistic feedback networks such as the Boltzmann machine can overcome the problem of local minima
- how a Boltzmann machine is trained
- how feedforward networks such as the radial basis function (RBF) network performs pattern recognition
- how an RBF network is trained
- how a probabilistic neural network (PNN) and a general regression neural network (GRNN) function

6.1 Introduction

Many neural networks have been developed that use some aspect of statistical theory. In this chapter both feedback and feedforward networks will be considered. Since the previous chapter was concerned with feedback networks, the first networks to be discussed are feedback networks that have been developed to overcome some of the inherent limitations of the basic associative memories.

6.2 Boltzmann machine

One of the problems with the Hopfield network is its tendency to settle at local energy minima rather than at the global minimum. Some of the local minima represent points in the 'energy landscape' where the stored patterns can be found. When an input vector is applied to the network, if it lies close to one of these minima, it will move towards that minimum because of the nature of the changes in the neurons, which always lower the energy in the network. So if a pattern is close to one of the stored patterns, it should produce that stored pattern at the output, once it has settled in the local minimum.

There may also be local minima which are not related to stored patterns. The effect of this is that the network can be shown a particular input pattern and, after a number of iterations, it will produce a stable output pattern. Later, the network could be shown the same input pattern and, because the system started in a different 'state', the output that it finally produces could be different, possibly even a completely new pattern.

The way to overcome this problem is to store the patterns at global minima and improve the method for finding minima. One way is to use probabilistic networks where occasional jumps to higher energy states are allowed rather than always moving towards lower energy. Then, if the network gets trapped in a local minimum, the occasional jump to a higher energy state will shake it out of that minimum and the search can continue to find the global minimum.

A physical analogy will now be given (Hinton, 1985) so that the method of overcoming this problem of being caught in a local minima can be appreciated. Imagine that the energy landscape is a bumpy surface where there are some shallow dents and some deeper ones, and the state of the network is a ball. If the surface is shaken up and down, the ball will leap about, in and out of the dents. If the ball is in one of the dents, the surface has to be shaken with enough vigour for it to be able to jump out of the dent into another one. If the energy put into the system is not enough, the ball will be stuck in the same dent.

We can see that the time spent in a particular dent will depend on the depth of that dent. The global minimum corresponds to the deepest dent of all. If the ball lands here, the maximum amount of energy will have to be put in to get the ball out again.

If the surface is shaken in a particular way for a long enough time, an equilibrium is reached where the ball is still moving about, but the probability of finding the ball in any one place is stable. This probability depends only on the height of the surface and does not depend on where the ball started. If the length of time is relatively short, the ball will be close to its initial position, so the probability of finding the ball in the region around the initial position will be exaggerated.

This physical analogy has a counterpart in thermodynamics (Kirkpatrick *et al.*, 1983). The energy being put into the system is thermal noise and the equilibrium state is called thermal equilibrium. An effective method of finding the global minimum is to mimic a process called simulated annealing. Annealing is the process used for hardening metal, where it is first heated to a high temperature, so the molecules have a lot of energy and therefore a lot of random motion. As the metal cools, the molecules settle into fixed positions with the tendency to look for minimum energy states.

In systems which are made of particles that can be in either of two states, the probability of the particle being in state S_a or state S_b follows the Boltzmann distribution:

$$\frac{p_a}{p_b} = e^{-(E_a - E_b)/T}$$

and since $p_a + p_b = 1$, it follows that:

$$p_a = \frac{1}{1 + e^{-(E_a - E_b)/T}}$$

where T is the temperature and E is the energy in each state.

A network has been proposed called the Boltzmann machine (Ackley *et al.*, 1985) in which the neurons produce 0 or 1 outputs depending on the value of the weighted sum, as before. The difference is that the outputs have a probability associated with them, so that the same weighted sum might produce a 1 output some of the time and a 0 output at others.

Figure 6.1 shows a typical layout of a Boltzmann machine. There need not be any sort of structure to the network, connections being made between any neurons. The connections are bidirectional and the weights are symmetric. So, for example, if the weight at the input to neuron i from neuron j is w_{ij}, then this equals the weight at the input to neuron j from neuron i, w_{ji}. The weights are real values, positive or negative.

Figure 6.1 A Boltzmann machine

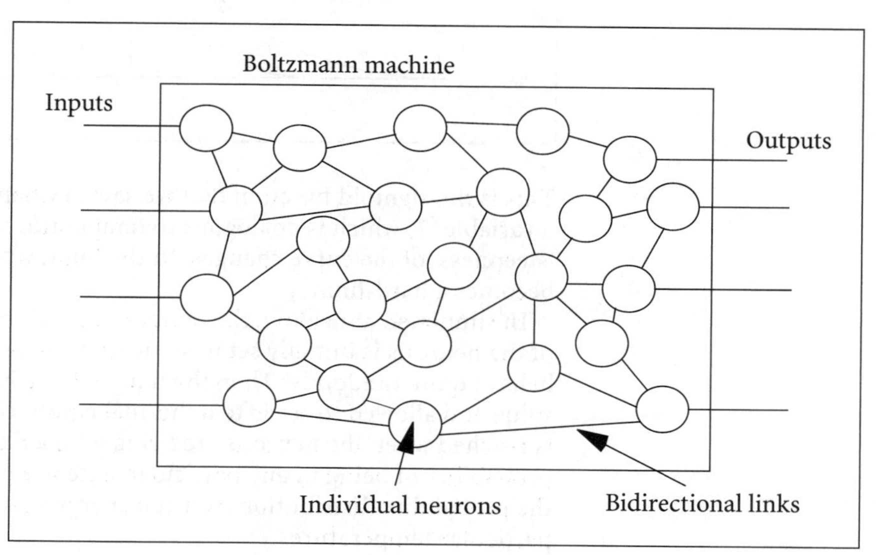

These neurons are designed to mimic the particles in a thermodynamic system, so their probability of being in either of two states also obeys the Boltzmann distribution, hence the name of the network.

From an earlier discussion in Chapter 5, the 'energy' associated with the two states of the neuron was said to be proportional to the weighted sum of the inputs. Thus the energy can be represented by:

$$\text{when } y = 1,\ E_1 = -\sum_{i=0}^{n} w_i x_i = -net$$

$$\text{and when } y = 0,\ E_0 = 0$$

Substituting into the equation for the probability of the output being 1, p_1:

$$p_1 = \frac{1}{1+e^{(E_1-E_0)/T}} = \frac{1}{1+e^{-\Sigma w_i x_i/T}} = \frac{1}{1+e^{-net/T}}$$

Figure 6.2 shows the distribution of the probability.

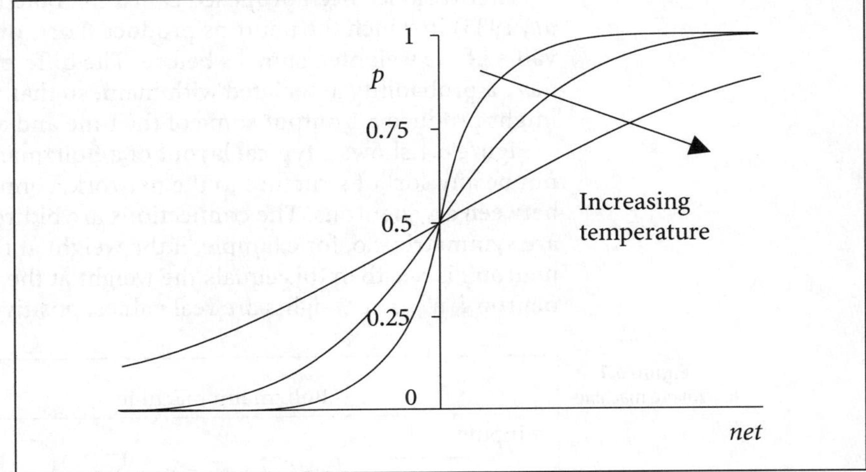

Figure 6.2 Sigmoid probability function

This is the sigmoid function that we saw in Chapter 3, with the addition of a variable, T, which is analogous to temperature. As T varies, the 'steepness' of the curve changes. In the limit, when $T = 0$, the function becomes a hard-limiter.

In simulated annealing, the temperature, T, in the probability functions of the neurons is initially set to some high value. As a result the neurons behave quite randomly. Then the temperature is brought down to a lower value and allowed to settle to a thermal equilibrium. Thermal equilibrium is reached when the neurons are firing with a fixed probability so that the probability of being in any particular state is equal to the value given in the probability distribution over the energy surface which is fixed for that particular temperature.

The temperature is continually lowered and allowed to settle to thermal equilibrium until the temperature is zero, when the global minimum energy state should be reached.

6.2.1 An example network

Figure 6.3 shows an example of a network (Hinton, 1985). There are seven neurons connected together in an arbitrary manner. Note that the connections between neurons are bidirectional, so the weight associated with each connection is written next to that connection. For example, the

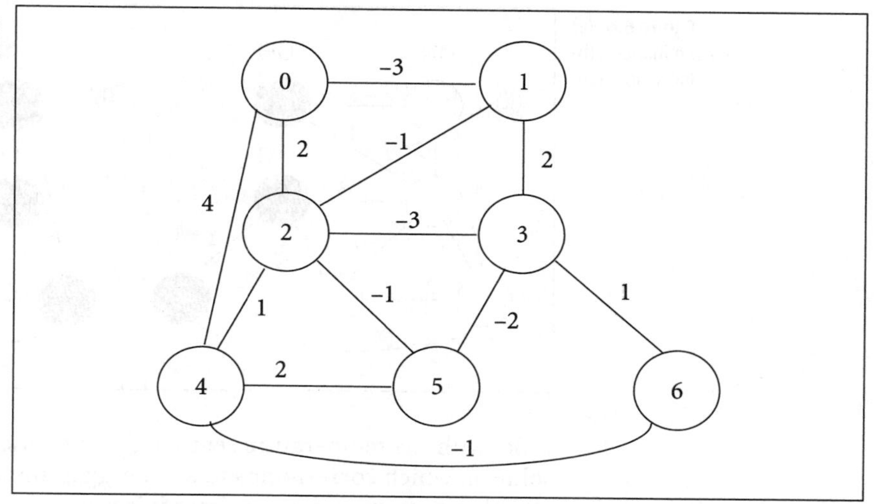

Figure 6.3 Example network

connection between neuron 0 and neuron 1 has a weight of –3. This means that $w_{01} = w_{10} = -3$.

It is assumed that for each neuron a weighted sum is calculated called *net*. In the first instance, assume that the temperature, T, is 0, so that each of the neurons behaves as a deterministic neuron which fires when the value of *net* is greater than zero. The value of *net* is calculated using all of the connections to the neuron.

$$\text{neuron fires } (=1) \text{ if } \sum_{i=0}^{n} w_i x_i > 0$$

$$\text{otherwise neuron does not fire } (=0)$$

Initially the neurons are set to some random state; then each neuron in turn has its *net* value calculated and its state altered accordingly. If all the neurons are initially set to 0, the network does not change.

In most cases the network settles to the state shown in Figure 6.4(a), where the open circles represent neurons that are firing. The energy in the network is the sum of all the energies of the individual neurons calculated using the equation given earlier as:

$$E = -\sum_{i=0}^{n} y_i \sum_{j=0}^{n} w_{ij} x_j = \sum_{i=0}^{n} y_i net_i$$

Since the inputs and outputs have values of 0 and 1, this equation reduces to finding the negative sum of the weights between active neurons and doubling it. The doubling factor, which comes about because of the symmetrical weight structure, $w_{ij} = w_{ji}$, will be ignored.

The energy in the network in Figure 6.4(a) turns out to be –8 and is the global minimum. An alternative stable solution is shown in Figure 6.4(b), where the energy is –3, and is therefore a local minimum.

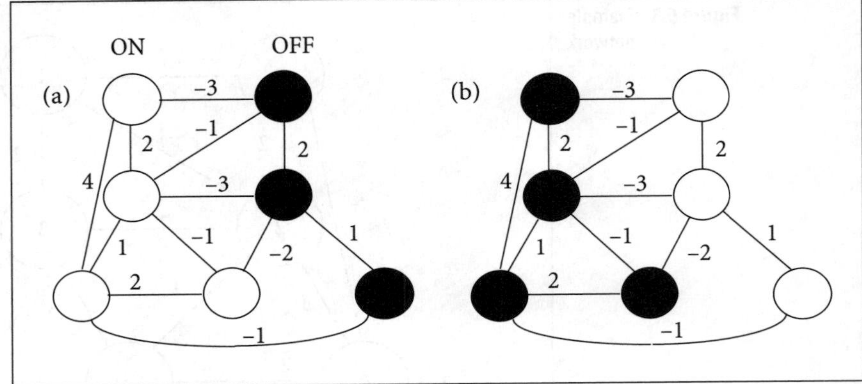

Figure 6.4 (a) Global minimum; (b) local minimum

So, with the temperature set to 0, the network converges to a stable solution which corresponds to an energy minima. Sometimes it is the global minimum and other times it is not.

Now raise the temperature. This means that the neurons calculate the value of *net*, and use it to set the probability of the output being a 1. One way of determining whether the output is going to be 0 or 1 is to calculate the probability, p, and then to generate a random number, r, between 0 and 1. The output is 1 if the value of $p > r$ and is 0 otherwise.

Again, the network starts in a random state with the temperature set to some value. If it is set to 1, the probability for each node firing is:

$$p_1 = \frac{1}{1+e^{-net}}$$

If the network is allowed to run, updating each of the neurons one at a time, after many iterations the neurons continue to fire with a given probability, but, measured over time, this probability is more or less constant. For example, after 1000 iterations, the neurons fire with a probability of:

Neuron	0	1	2	3	4	5	6
Probability	0.978	0.055	0.845	0.087	0.987	0.699	0.292

In these probabilistic neurons, convergence to a stable solution means something quite different to settling at a constant value. Here, the network converges to a solution which is changing but with a fixed probability. As the temperature gets lower, the probabilities get closer to 0 and 1, so that in the limit, when $T = 0$, the probabilities are:

Neuron	0	1	2	3	4	5	6
Probability	1	0	1	0	1	1	0

If the probability of producing an output of 1 is 1.0, the output is permanently 1. Similarly, if the probability of producing an output of 1 is

0.0, the output is never 1, which means that it is always 0, since the output can only be either 0 or 1. Therefore the probabilities in this solution can be interpreted as the values of the output, which corresponds to a state that was shown earlier to be the global minimum.

So if the network starts at a high temperature and gradually reduces it, the network should end up in the global minimum rather than a local minimum. To test this, the network was started from the local minimum that was found earlier, with the temperature set to a relatively low value, 0.04.

Neuron	0	1	2	3	4	5	6
Output	0	1	0	1	0	0	1

For 20 iterations the network stayed in this state; then, on the next iteration, neuron 5 produced a 1 output instead of a 0. The network continued to change until after only 6 iterations it had reached the global minimum solution:

Neuron	0	1	2	3	4	5	6
Output	1	0	1	0	1	1	0

The conclusion that can be drawn is that the addition of probabilistic outputs helps to overcome the problem of getting trapped in local minima by occasionally lifting the network to a higher energy state. As the temperature gets lower, this becomes less frequent and, ultimately, the network settles at $T = 0$ in the global minimum.

6.2.2 Training the Boltzmann machine

Training the Boltzmann machine consists of a combination of simulated annealing and weight adjustment. Some of the neurons are chosen as inputs and some others are chosen as outputs. Training has two distinct phases:

Phase 1
The input pattern and the corresponding output pattern are imposed on the input and output neurons. This is described as clamping the neurons, so that their activity is fixed irrespective of their own internal *net* values. The temperature is set to some arbitrary high value and the network is allowed to settle into thermal equilibrium. It is then gradually cooled, by lowering the temperature by a small amount and then leaving the network to settle in thermal equilibrium. When the temperature reaches a low value the network should be in a stable state which should in principle be the global energy minimum for that set of weights.

Phase 2
The clamp on the output is removed and the process of simulated annealing repeated. The network will again settle into what is hoped to be

the global minimum, but because the outputs are probably different, this minimum will be different from the minimum found in Phase 1.

Since the network starts off with arbitrary weight values there will almost certainly be conflicts in Phase 1. Conflicts arise when the *net* value of one of the output neurons suggests that the neuron should be in a particular state, but it is clamped in the opposite state. The energy will therefore be greater and the global minimum will be higher than it could be.

In Phase 2 the neurons are no longer clamped, so the output neurons settle at the values decided by the value of *net*. Only when the weights have been adjusted so that no more conflicts arise will the state of the network in both phases be the same.

The weights have to be adjusted in some way. The outputs in Phase 1 could be compared to the outputs in Phase 2 and an error produced, but since the network has complex interconnections between neurons, no algorithm exists for back-propagating the errors.

The alternative method (Hinton, 1985) is to adjust the weights while the temperature is still greater than 0, when the network is still behaving in a probabilistic manner. Small amounts are added or subtracted to the weights in each neuron so that the probabilities of two neurons firing together are the same in both phases.

In Phase 1, the network is allowed to settle to its thermal equilibrium with $T > 0$. The network is then run for a fixed number of iterations. During each iteration, the weights in each neuron are adjusted by a small amount, δ, if the neuron containing the weights and the corresponding inputs to that neuron are both 1.

$$w_{ij} = w_{ij} + \delta \text{ if } x_i = 1 \text{ and } x_j = 1$$

The output pattern is then removed, and the system allowed to settle into equilibrium again. The input is still clamped, but the output is free to settle to any value. The network is allowed to run for the same fixed number of iterations again. This time, during each iteration, the weights are adjusted by an amount equal to $-\delta$ if the output of the neuron containing the weights and the corresponding inputs to that neuron are both 1.

$$w_{ij} = w_{ij} - \delta \text{ if } x_i = 1 \text{ and } x_j = 1$$

This process of clamping and unclamping is repeated for the other input/output pairs in the training set until the changes to the weights fall below a fixed tolerance.

Strictly speaking, for a gradient descent method each weight should be adjusted according to the formula:

$$\Delta w = k(p_{ij} - p'_{ij})$$

where p_{ij} is the average probability of two neurons, i and j, being on when the network is clamped, and p'_{ij} is the probability of them being on when it is not clamped, and k is a scaling factor (Ackley *et al.*, 1985). However, simply adding and subtracting a value δ according to the learning rule given above should produce the same result.

For example, assume that $p_{ij} = 0.5$ and $p'_{ij} = 0.4$. Then at each iteration, a value of $0.1k$ should be added to the weights. Over 10 iterations this equals k.

If, instead, δ is added when both neurons are firing during the clamped stage, and subtracted when both are firing during the unclamped stage, the result, on average, will be to add δ five times out of 10 and to subtract δ four times out of 10. Hence, a resultant addition of δ over 10 iterations. So, in both cases, the end result is that a small value is added to the weight, either k or δ.

The temperature is then lowered, and the whole process repeated. This continues until the temperature reaches zero, by which time the weights should have been adjusted to produce a global minimum energy for that pattern.

There are a number of problems associated with Boltzmann machines (Rumelhart and McClelland, 1986). Decisions about how much to adjust the weights, how long you have to collect statistics to calculate the probabilities, how many weights you change at a time, and how you adjust the temperature during the simulated annealing. To these could be added the question of how do you know when the network has reached thermal equilibrium? However, the main drawback of the Boltzmann machine is the length of time required for it to learn. To quote Hinton (Hinton, 1985):

> Our current simulations are slow for three reasons: It is inefficient to simulate parallel networks with serial machines, it takes many decisions by each neuron before a big network approaches equilibrium, and it takes an inordinate number of I/O pairs before a network can figure out what to represent with its internal neurons. Better hardware might solve the first problem, but more theoretical progress is needed on the other two. Only then will we be able to apply this kind of learning network to more realistic problems.

6.3 Cauchy machine

The Cauchy machine (Szu and Hartley, 1987) is very similar to the Boltzmann machine but a different probability function is used.

Boltzmann	Cauchy
$\frac{p_a}{p_b} = e^{-(E_a - E_b)/T}$	$\frac{p_a}{p_b} = \frac{T}{T + (E_a - E_b)^2}$

It is claimed that this change in the probability function allows jumps to higher energy states more often so that 'fast annealing' can take place.

6.4 PLN

In Chapter 4 logic nodes were briefly considered. A fully connected canonical network (Aleksander, 1989b) consists of a network of either RAMs or PLNs arranged so that if there are n nodes, then each node has $n - 1$ inputs connected to the output of every other node except its own. Figure 6.5 shows such a network for $n = 3$.

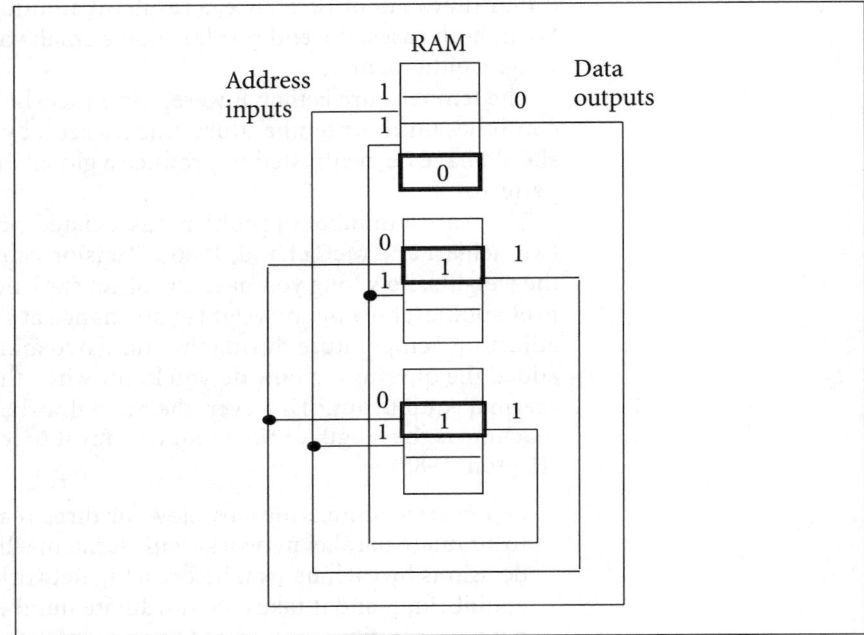

Figure 6.5 Fully connected canonical network for $n = 3$

If the nodes are RAMs, then the network can be trained so that its behaviour is very similar to the Hopfield network. The aim is to store patterns in a stable state. This is achieved by placing the required pattern on the outputs of each of the nodes which addresses a particular location in each node. The same pattern is then stored at those locations. Figure 6.5 shows the case for the pattern 011.

It is usually assumed that the initial contents of the RAMs are random. Despite the very simple nature of the training, it suffers from the same drawbacks as the Hopfield network, which is the creation of false minima and hence undesirable stable states. This can be overcome in a similar way to the Boltzmann machine, namely the introduction of probability of firing.

RAMs can be made probabilistic by interpreting the q-bit number stored in a RAM at a particular address as a probability. So if there are three bits, and the number stored is 010, then this is interpreted as a probability of 2/7. A special case exists when the RAM only stores three possible numbers, and these are interpreted as 0, 1 and d which stands for 'don't know'. When the stored data is *d*, the output produces a 0 or a 1 with equal probability.

Before training, all of the locations in the RAMs are set to *d*. So the probability of being in any of the states is equal. In the example already described, where there are three nodes, there are eight possible states, so the probability of being in any one is 0.125.

Training is the same as before, namely, placing the required pattern on the output and also placing it in the locations that are addressed by the pattern. This means that the *d*s get changed to 0s and 1s at those locations, but stay as *d* in the remaining locations.

An interesting comparison can be made with the Boltzmann machine if energy, *E*, is found. The relationship between the probability of being in a particular state and the energy is defined using the Boltzmann relationship:

$$p = \frac{1}{1+e^{-kE}}$$

where *k* is a constant.

So, if the probability of being in a particular state at any time is known, the energy can be calculated. However, some changes have to be made. The first is that, in the Boltzmann machine, a high value for energy implies a high probability, whereas in the PLN network a high probability implies a low energy. So the value of $1 - p$ is used instead.

Next, it is assumed that $k = 1$ for convenience. Finally, it sometimes happens that the probability of being in a particular state is 0 or 1, which implies an energy of $\pm\infty$. To avoid this, the energy is scaled to be within the arbitrarily chosen limits of ±5. So the values for energy, *E*, can be derived from the known probabilities, p', and will lie in the range:

$$p' = 0 \qquad E = -5$$

$$p' = 0.5 \qquad E = 0$$

$$p' = 1 \qquad E = +5$$

Thus the new formula is:

$$0.00667 + 0.98666p' = \frac{1}{1+e^{E}}$$

$$E = \ln(0.00667 + 0.98666p') - \ln(0.99333 - 0.98666p)$$

Using these formulae it is possible to show that the stable states exist in energy minima. Furthermore, because probabilistic nodes are used, jumps to higher energy states can occur so it is possible to escape from false local minima. These are desirable features which compare very favourably with the Boltzmann machine.

A more detailed analysis can be found in Aleksander's book on neural computing architectures (Aleksander, 1989a).

6.5 Radial basis function networks

The previous sections were all concerned with feedback networks. Statistics is also being used in feedforward networks, and one of the most important of these is the radial basis function (RBF) network (Broomhead and Lowe, 1988). This is becoming an increasingly popular neural network with diverse applications and is probably the main rival to the multi-layered perceptron.

Much of the inspiration for RBF networks has come from traditional statistical pattern classification techniques, which are essentially getting a new lease of life as a form of neural network. However, by including RBFs in the general category of neural networks these techniques, which would only have been known to the few, have become widely used.

6.5.1 Architecture

The basic architecture for an RBF is a three-layer network, as shown in Figure 6.6. The input layer is simply a fan-out layer and does no processing. The second or hidden layer performs a non-linear mapping from the input space into a (usually) higher-dimensional space in which the patterns become linearly separable. The final layer therefore performs a simple weighted sum with a linear output. If the RBF network is used for function approximation then this output is fine. However, if pattern classification is required, then a hard-limiter or sigmoid function could be placed on the output neurons to give 0 or 1 output values.

Figure 6.6 Radial basis function network

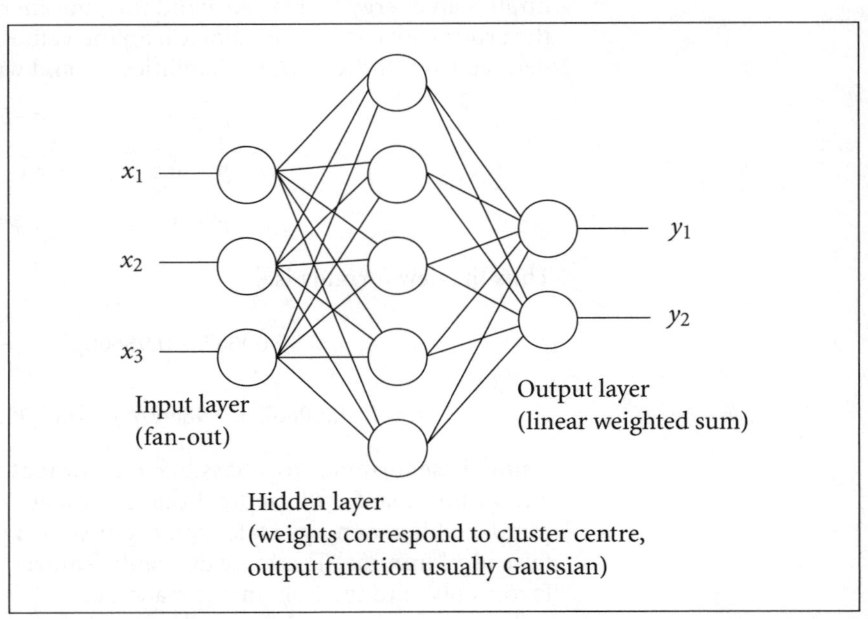

The unique feature of the RBF network is the process performed in the hidden layer. The idea is that the patterns in the input space form clusters. If the centres of these clusters are known, then the distance from the cluster centre can be measured. Furthermore, this distance measure is made non-linear, so that if a pattern is in an area that is close to a cluster centre it gives a value close to 1. Beyond this area, the value drops dramatically. The notion is that this area is radially symmetrical around the cluster centre, so that the non-linear function becomes known as the radial basis function. The most commonly used radial basis function is:

$$\phi(r) = \exp\left(-\frac{r^2}{2\sigma^2}\right)$$

In an RBF network, r is the distance from the cluster centre. The equation represents a Gaussian bell-shaped curve, as shown in Figure 6.7.

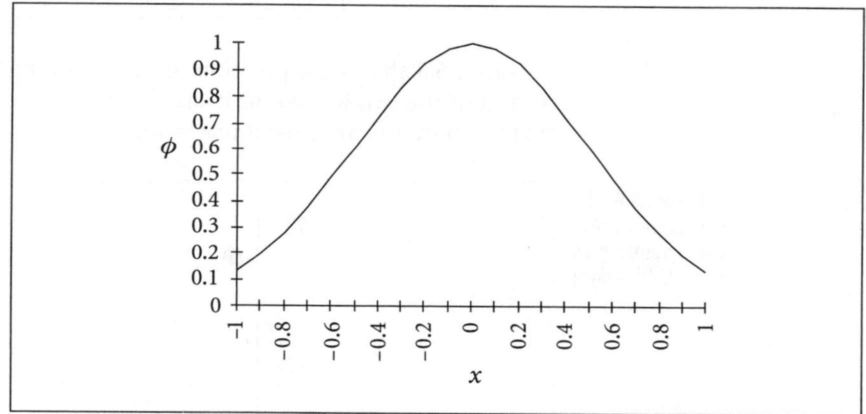

Figure 6.7 A Gaussian centred at 0 with $\sigma = 0.5$

The distance measured from the cluster centre is usually the Euclidean distance. For each neuron in the hidden layer, the weights represent the coordinates of the centre of the cluster. Therefore, when that neuron receives an input pattern, X, the distance is found using the following equation:

$$r_j = \sqrt{\sum_{i=1}^{n}(x_i - w_{ij})^2}$$

So the output of neuron j in the hidden layer is:

$$\phi_j = \exp\left(-\frac{\Sigma_{i=1}^{n}(x_i - w_{ij})^2}{2\sigma^2}\right)$$

The variable σ defines the width or radius of the bell shape and is something that has to be determined empirically. When the distance from the centre of the Gaussian reaches σ, the output drops from 1 to 0.6.

An often quoted example which shows how the RBF network can handle a non-linearly separable function is the exclusive-or problem

(Haykin, 1999). One solution has two inputs, two hidden units and one output. The centres for the two hidden units are set at c1 = 0,0 and c2 = 1,1, and the value of radius σ is chosen such that $2\sigma^2 = 1$. When all four examples of input patterns are shown to the network, the outputs of the two hidden units are shown in the following table. The inputs are x, the distances from the centres squared are r, and the outputs from the hidden units are ϕ.

x_1	x_2	r_1	r_2	ϕ_1	ϕ_2
0	0	0	2	1	0.1
0	1	1	1	0.4	0.4
1	0	1	1	0.4	0.4
1	1	2	0	0.1	1

Figure 6.8 shows the position of the four input patterns using the output of the two hidden units as the axes on the graph. It can be seen that the patterns are now linearly separable.

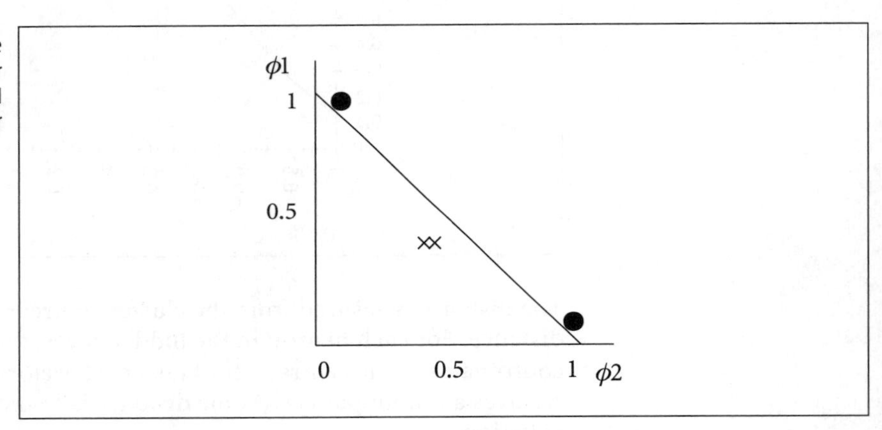

Figure 6.8 The input patterns after being transformed by the hidden layer

This demonstrates the power of transforming from one domain to another using an RBF network. However, the centres were chosen carefully to show this result. The methods generally adopted for learning in an RBF network would find it impossible to arrive at those centre values. In the next section the learning methods that are usually adopted will be described.

6.5.2 Training an RBF network

Hidden layer

The hidden layer in an RBF network has units which have weights that correspond to the vector representation of the centre of a cluster. These weights are found either using a traditional clustering algorithm such as the *k*-means algorithm (Duda and Hart, 1973), or adaptively using essentially the Kohonen algorithm that will be described in the next chapter. In either case, the training is unsupervised, but the number of clusters that you expect, k, is set in advance. The algorithms then find the best fit to these clusters. Kohonen learning will be described in the next chapter, so the k-means algorithm will be briefly outlined.

Initially, k points in the pattern space are randomly set. Then for each item of data in the training set, the distances are found from all of the k centres. The closest centre is chosen for each item of data. This is the initial classification, so all items of data will be assigned a class from 1 to k. Then, for all data which has been found to be class 1, the average or mean values are found for each of the coordinates. These become the new values for the centre corresponding to class 1. This is repeated for all data found to be in class 2, then class 3 and so on until class k is dealt with. We now have k new centres. The process of measuring the distance between the centres and each item of data and re-classifying the data is repeated until there is no further change. The sum of the distances can be monitored and the process halts when the total distance no longer falls.

The alternative is to use an adaptive k-means algorithm which is the same as Kohonen learning. Input patterns are presented to all of the cluster centres one at a time, and the cluster centres adjusted after each one. The cluster centre that is nearest to the input data wins, and is shifted slightly towards the new data. This has the advantage that you do not have to store all of the training data, so it can be done online.

Having found the cluster centres using one or other of these methods, the next step is to determine the radius of the Gaussian curves. This is usually done using the P-nearest neighbour algorithm. A number P is chosen, and for each centre, the P nearest centres are found. The root mean squared distance between the current cluster centre and its P nearest neighbours is calculated, and this is the value chosen for σ. So, if the current cluster centre is c_j, the value is:

$$\sigma_j = \sqrt{\frac{1}{P}\sum_{i=1}^{P}(c_k - c_i)^2}$$

A typical value for P is 2, in which case σ is set to be the average distance from the two nearest neighbouring cluster centres.

Using this method for training the hidden layer, the exclusive-or function can be implemented using a minimum of four hidden units. If more than four units are used, the additional units duplicate the centres and therefore do not contribute any further discrimination to the network. So, assuming four neurons in the hidden layer, each unit is centred on one of the four input patterns, namely 00, 01, 10 and 11. The P-nearest neighbour algorithm with P set to 2 is used to find the size if the radii. In each of the neurons, the distances to the other three neurons are 1, 1 and 1.414, so the two nearest cluster centres are at a distance of 1. Using the mean squared distance as the radii gives each neuron a radius of 1.

Using these values for the centres and radius, if each of the four input patterns is presented to the network the output of the hidden layer would be:

Input	Neuron 1	Neuron 2	Neuron 3	Neuron 4
00	0.6	0.4	1.0	0.6
01	0.4	0.6	0.6	1.0
10	1.0	0.6	0.6	0.4
11	0.6	1.0	0.4	0.6

Output layer
Having trained the hidden layer with some unsupervised learning, the final step is to train the output layer using a standard gradient descent technique, such as the LMS algorithm used for ADALINE. In the example of the exclusive-or function given above a suitable set of weights would be +1, –1, –1 and +1. With these weights the value of *net* and the output are:

Input	Neuron 1	Neuron 2	Neuron 3	Neuron 4	*net*	Output
00	0.6	0.4	1.0	0.6	–0.2	0
01	0.4	0.6	0.6	1.0	0.2	1
10	1.0	0.6	0.6	0.4	0.2	1
11	0.6	1.0	0.4	0.6	–0.2	0

6.5.3 Advantages of an RBF

Many advantages are claimed for RBF networks over multi-layer perceptrons (MLP). It is said that an RBF trains faster than an MLP and that it produces better decision boundaries. The evidence for this is given by Leonard and Kramer (1991) who compare RBF and MLP networks for a specific problem. Another advantage that is claimed is that the hidden layer is easier to interpret than the hidden layer in an MLP (Leonard *et al.*, 1992). There is some evidence for this, as we shall see in the next section, where the hidden layer in an RBF is interpreted as a probability density function. Some of the disadvantages that are claimed are that an MLP gives better distributed representation.

6.6 Probabilistic neural network (PNN)

Earlier it was said that RBF networks were developed from traditional statistical pattern classification techniques. This is even more true of the probabilistic neural network developed by Specht (1990). We have already seen in this chapter how probabilistic networks are used in feedback networks, where the states that the machine enters are probabilistically determined. The networks described here are feedforward networks that try to approximate the underlying probability density function of the patterns being classified.

In RBF networks it is assumed that Gaussian functions make good approximations to the cluster distribution in the pattern space. This is quite an over-simplification, and quite unlikely to be true. However, it has been shown by Parzen (1962) that the probability density function of a particular class in a pattern space can be approximated by a sum of kernel functions, and one possible kernel is the Gaussian function. This means that, with only minor alterations, RBF networks can be used to approximate the probability density functions for each class of pattern, which in turn can be used to classify the patterns with a high probability of success.

Parzen showed that if a kernel function is centred on each piece of data from one class in a training set, then the sum of the kernel functions is a good approximation to the overall probability density of that class. If Gaussians are used, then they are centred over each data point from a class. The probability density function for a class is approximated using the following equation:

$$\mathrm{pdf}_k(x) = \left(\frac{1.0}{2\pi^{n/2}\sigma^n}\right)\left(\frac{1.0}{P_k}\right)\sum_{j=1}^{P_k} \mathrm{e}^{(-(x-x_{kj})^2/(2\sigma^2))}$$

The value of P_k is the number of data in class k, and n is the number of inputs. The term x_{kj} represents the centre of a Gaussian function and corresponds to item j in the set of data belonging to class k. Although this looks quite complex, it only means that the sum of the Gaussians is taken and averaged, and then a weighting factor applied. This weighting factor (in the first brackets) consists of constant terms and the radius to the power of n.

This only leaves the decision about the value of the radius, σ. Choosing a value that is too large results in over-generalisation, and choosing a value too small results in over-fitting. As an example, Figure 6.9 shows two classes of data defined by a single variable. The data for class A are at 0, 1, 3 and 6 and the data for class B are at 8, 8.1, 8.2, 8.5 and 9. A Gaussian curve with a radius of 1.5 is drawn, centred at each data point, and the sum taken for each class to get the approximate probability density functions for each class. The pdf values have not been normalised to emphasise their values relative to each other.

The radius was chosen by trial and error to give the smooth functions shown in the figure. Now if a new sample of data was to be classified and

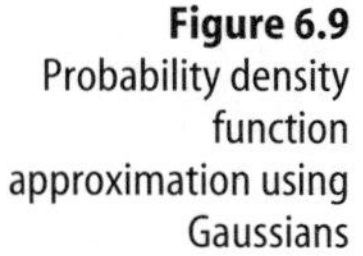

Figure 6.9 Probability density function approximation using Gaussians

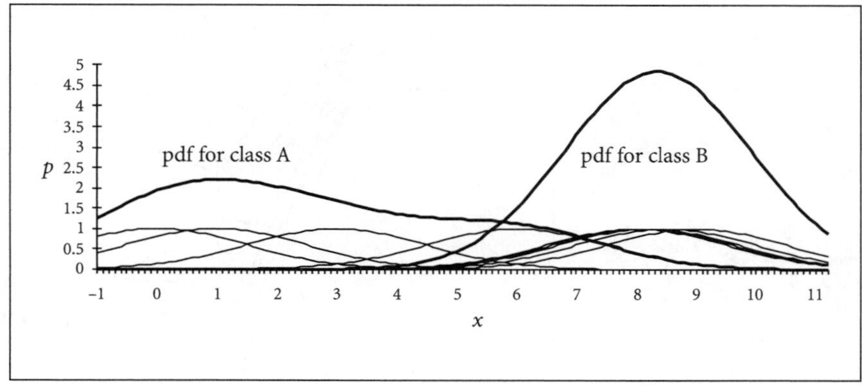

was found to have a value of 6.5, the nearest data is at 6 which is class A. However, the pdf for class A at 6.5 gives a value just over 1, whereas the pdf for class B gives a value closer to 2. Had the values been normalised they would lie between 0 and 1. However, what this shows is that there is a higher probability that the new data belongs to class B than class A even though the nearest data belongs to A. This is because of the density of the data. The data belonging to B are bunched tightly together, giving a higher density than the data in A, which are more spread out.

Parzen showed that the value of σ should be dependent on the number of patterns in a class. One function that is often used is:

$$\sigma = aP_k^{-b}$$

The value of b is a constant between 0 and 1.

A probabilistic network uses these ideas to produce a pattern classifier. A three-layer network like the RBF network is used with each unit in the hidden layer centred on an individual item of training data. The value of the radius, σ, is given by the formula above, with arbitrary values chosen for a and b.

Each unit in the output layer has weights of 1, and a linear output function, so this layer simply adds all of the outputs from the hidden layer that correspond to data from the same class together. This output represents the probability that the input data belongs to the class represented by that unit. The final decision as to what class the data belongs to is simply the unit in the output layer with the largest value. In some networks an additional layer is included that makes this decision as a winner takes all circuit. Figure 6.10 shows a network that finds three

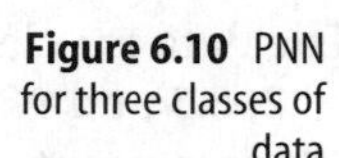

Figure 6.10 PNN for three classes of data

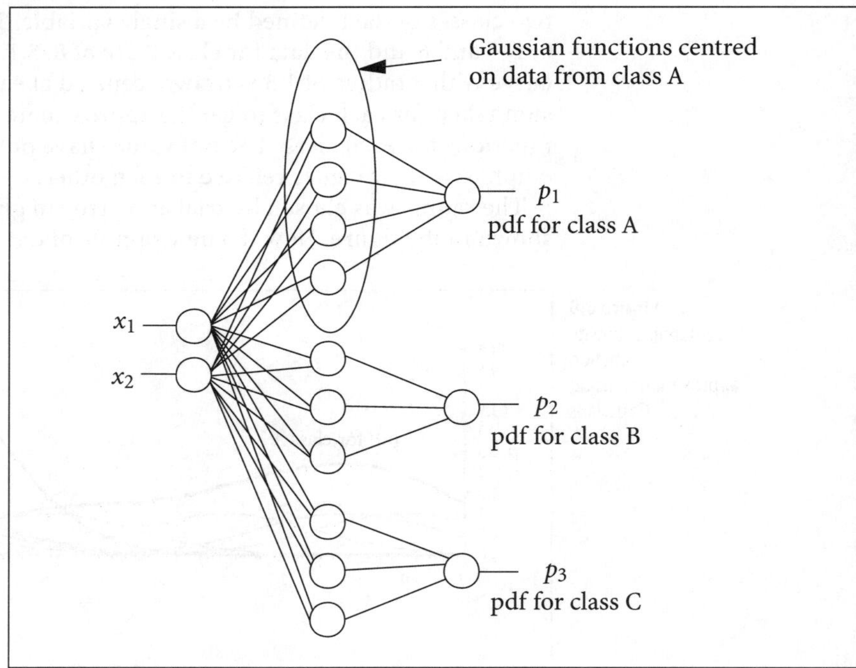

classes of data. There are two input variables, and for the three classes of data there are four samples for class A, three samples for class B and three samples for class C; hence the number of neurons in the hidden layer.

The advantage of the PNN is that there is no training. The values for the weights in the hidden units (i.e. the centres of the Gaussian functions) are just the data values themselves. In the output layer the weights are all 1 and the radius, σ, is determined by the number of samples in each class in the training set. One disadvantage is the number of units needed, which matches the amount of data. For problems with sparse data, however, this is a good solution. If the amount of data is very large, then some clustering can be done to reduce the number of units needed in the hidden layer.

6.7 General regression neural network (GRNN)

The general regression neural network was also developed by Donald Specht (1991) for system modelling and identification and can be thought of as a generalisation of the probabilistic neural network. The PNN is specifically used for pattern classification, whereas the GRNN has broader applications.

The idea behind the GRNN is that you can approximate any function given a set of input and output data pairs. Assuming that a function has n inputs, x_1 to x_n and one output, y, the expected mean value of the output given a particular input x can be found using the following equation from probability theory:

$$\bar{y}(x) = \frac{\int_{-\infty}^{\infty} y \,\mathrm{pdf}(x, y) \mathrm{d}y}{\mathrm{pdf}(x)}$$

This equation contains the function pdf(x,y) which is a joint probability density function, i.e. the probability of the output being y and the input being x. In the same way the probability density functions were approximated by the sum of Gaussians in the PNN, so the conditional probability functions can also be approximated. This is done by placing the centre of a Gaussian function over the input data in a training set, and multiplying the Gaussian by the value of the corresponding output. After some manipulation, the equation given above can be approximated by the following equation:

$$\hat{y} = \frac{\Sigma_{p=1}^{P} y_p \exp(-d_p^2 / 2\sigma^2)}{\Sigma_{p=1}^{P} \exp(-d_p^2 / 2\sigma^2)}$$

In this equation it is assumed that there are P items of data in the training set. Each item consists of a training pair containing the input vector, x_{1p}, x_{2p} to x_{np} and the corresponding output value y_p. The value of d_p is the distance between the current input vector and the pth input vector in the training data. For a given input vector, the distance from it and all of the input vectors in the training set is found and passed through a Gaussian function centred on the input vectors from the training set. For the bottom of the equation, the outputs from these Gaussian functions are added together to get a total. On the top of the equation, the outputs from the Gaussian functions are weighted by the corresponding value of y from the training set before adding them together.

The radii of the Gaussian functions is again an issue, and as before they are found using the following equation, where b has a value between 0 and 1.

$$\sigma = aP^{-b/n}$$

From the point of view of a neural network architecture, it is very similar to the PNN except that the weights in the output layer are not set to 1. Instead they are set to the corresponding values of the output y in the training set. In addition, the sum of the outputs from the Gaussian layer has to be calculated, so that the final output can be divided by the sum. The architecture is shown in Figure 6.11 for a single output function of two variables with a set of 10 data points.

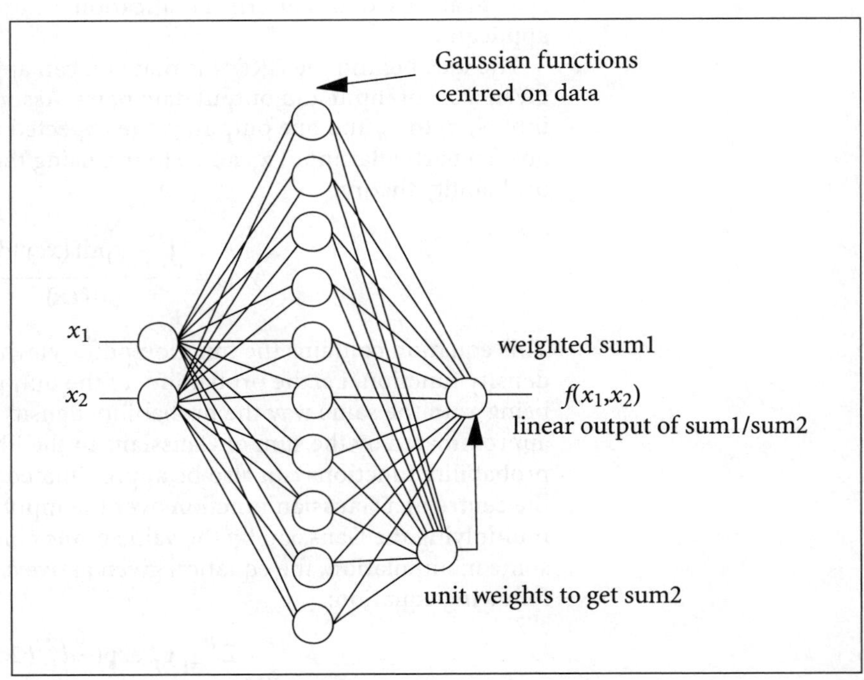

Figure 6.11 Architecture of a GRNN

If the output is a vector, then each element of the output vector has its own GRNN, as described above. As with the PNN, the GRNN is very memory-intensive, as it has to store all of the training data. However, it learns immediately and is very useful when the data is sparse. In some versions of the GRNN, the data is clustered first of all to reduce the number of units needed in the hidden layer.

CHAPTER SUMMARY

This chapter has dealt with statistical feedback networks such as the Boltzmann machine and the PLN. In both cases, the aim of using probabilistic outputs is to be able to avoid local minima in the energy landscape. The input pattern is the initial state in a sequence ending with the desired output pattern. It is hoped that by training using simulated annealing, this final state will be stored in the global energy minimum.

Thus statistical feedback networks overcome the problem of local minima. However, the inherent difficulties of constructing these networks so that they can operate in a reasonable time has proved to be too difficult, with the result that these networks are rarely applied to anything but simple problems.

On the other hand, the statistical feedforward networks, such as the radial basis function network, have become very popular, and are serious rivals to the multi-layered perceptron. Their success is likely to be due to the fact they are essentially well-tried statistical techniques being presented as neural networks.

The learning mechanisms in statistical neural networks are not biologically plausible, with the result that these networks have not been taken up by those researchers who insist on biological analogies. In the next chapter, on self-organising networks, some of the networks that will be considered are more biologically plausible, at least in the sense that they were either inspired from biology, in the case of the Kohonen networks, or are attempting to explain some biological phenomena in the case of the ART network.

SELF-TEST QUESTIONS

1. For the Boltzmann machine shown in Figure 6.3, what is the energy in the network when all of the neurons are firing, i.e. all have a value of 1?
2. A set of data belong to two classes – A and B. The data in A are 0, 1, 3 and 6, and the data in B are 8.0, 8.1, 8.2, 8.5 and 9.0. Use the k-means algorithm to find the cluster centres assuming that $k = 2$ and that the initial random values are 4.0 and 10.0.
3. With the centre values found in Question 2, what would the values be for the radii of the two neurons in the hidden layer of an RBF when the P-nearest neighbour algorithm is used, assuming in this case that P is 1, since there are only two classes?
4. With the centre value from Question 2 and the radii from Question 3, what would the values of the output of the hidden units in an RBF be for all of the input data?

SELF-TEST ANSWERS

1 If all the neurons are firing, then all of their outputs are 1. To find the energy you first have to find the weighted sum, *net*, for each neuron. Since all the inputs will be 1, the weighted sum is the sum of all of the weights on the inputs connected to each neuron. Starting at neuron 0, there are three inputs weighted 4, 2 and –3, so the value of *net* is 4 + 2 –3 = 3. The energy in this neuron is found by multiplying *net* with the output, *y*, and taking the negative value. Since all *y*s are 1, the energy in this neuron is –3.

Neuron	*net*	Energy
Neuron 0	4 + 2 – 3 = 3	–3
Neuron 1	–3 – 1 + 2 = –2	2
Neuron 2	1 + 2 – 1 – 3 – 1 = –2	2
Neuron 3	–3 – 2 + 2 + 1 = –2	2
Neuron 4	4 + 1 + 2 – 1 = 6	–6
Neuron 5	2 – 1 – 1 – 2 = –2	2
Neuron 6	1 – 1 = 0	0

The total energy is found by adding up all the individual energies to give a value of –1. This value is higher than the two values given in the text for stable states.

2 The initial classification is:

Data	Distance from C1	Distance from C2	Chosen centre
0.0	4.0	10.0	C1
1.0	3.0	9.0	C1
3.0	1.0	7.0	C1
6.0	2.0	4.0	C1
8.0	4.0	2.0	C2
8.1	4.1	1.9	C2
8.2	4.2	1.8	C2
8.5	4.5	1.5	C2
9.0	7.0	1.0	C2

From this initial classification, the average value is taken for each class to get new centres. For C1 the average is 2.5, and for C2 the average is 8.36 which will be rounded to 8.4. Repeat with these new centre values to get:

Data	Distance from C1	Distance from C2	Chosen centre
0.0	2.5	8.4	C1
1.0	1.5	7.4	C1
3.0	0.5	5.4	C1
6.0	3.5	2.4	C2
8.0	5.5	0.4	C2
8.1	5.6	0.3	C2
8.2	5.7	0.2	C2
8.5	6.0	0.3	C2
9.0	6.5	0.6	C2

The data at 6.0 gets classified as class 2 this time. So now when we find the average values the new centres are at 2.0 and 7.96, which will be rounded to 8.0. Further iterations do not change these classifications, so these are the final centre values.

3 The distance between the clusters is 8.0 – 2.0 = 6.0. Since there is only one value, it is the one chosen for the radii of both neurons.

4 The output for both neurons in the hidden layer is given in the following table.

$$\text{Neuron } 1 = e^{-(x-2.0)^2/72.0}$$

$$\text{Neuron } 2 = e^{-(x-8.0)^2/72.0}$$

Data	Neuron 1	Neuron 2
0	0.945959	0.411112
1	0.986207	0.506336
3	0.986207	0.706648
6	0.800737	0.945959
8	0.606531	1.000000
8.1	0.596423	0.999861
8.2	0.586320	0.999445
8.5	0.556101	0.996534
9	0.506336	0.986207

The final step in the design of an RBF would be to use the LMS algorithm to train the output layer to distinguish the two classes. Something worth pointing out is the problem that has been created by using unsupervised learning to get the cluster centres. The datum at 6.0 belongs to class A but has been clustered with class B. Although it is still possible to find a set of weights that will separate the two classes of data, it would have been much easier if the datum at 6.0 had been clustered with the rest of that class.

CHAPTER 7

Self-organising networks

CHAPTER OVERVIEW

This chapter gives a description of neural networks that use unsupervised learning and therefore exhibit self-organisation. It explains:

- how instar and outstar networks respond and how they are trained
- how inputs and weights can be represented by vectors
- what is meant by competitive learning
- how self-organisation is achieved using example networks such as the counter-propagation and Kohonen networks, ART and the neocognitron

7.1 Introduction

Another variation on the neural network are systems which are said to be self-organising networks. What this means is that the systems are trained by showing examples of patterns that are to be classified, and the network is allowed to produce its own output code for the classification.

In all the previous networks that have been discussed, the output was provided by the user during training (the only exception to this being the hidden units in the RBF). This is known as supervised training. In self-organising networks the training can be supervised or unsupervised. The advantage of unsupervised learning is that the network finds its own energy minima and therefore tends to be more efficient in terms of the number of patterns that it can accurately store and recall.

Obviously, it is up to the user to interpret the output. During training, input patterns are shown, and when the corresponding output patterns are produced, the user knows that that code corresponds to the class which contains the input pattern.

In self-organising networks four properties are required:

- The weights in the neurons should be representative of a class of patterns. So each neuron represents a different class.

- Input patterns are presented to all of the neurons, and each neuron produces an output. The value of the output of each neuron is used as a measure of the match between the input pattern and the pattern stored in the neuron.
- A competitive learning strategy which selects the neuron with the largest response.
- A method of reinforcing the largest response.

The following sections look at different networks to see how they implement these properties. The inspiration for many of these networks came from biology. They have been developed either to model some biological function, particularly in cognition, or in response to the demand for biological plausibility in neural networks.

7.2 Instar and outstar networks

Instar networks and outstar networks (Carpenter, 1989) form the basis of a number of networks including the counter-propagation network (Hecht-Nielsen, 1987) and the ART networks (Carpenter and Grossberg, 1988). The neurons in the instar network, shown in Figure 7.1, are essentially the same as the neurons in other networks that have been seen in the previous chapters. They take inputs, x_i, and produce a weighted sum called net_j:

$$net_j = \sum_{i=0}^{n} w_{ij} x_i$$

This weighted sum is the output of the neuron, which means that there is no non-linear output function in the neurons.

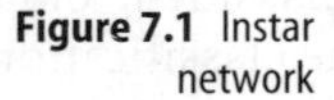
Figure 7.1 Instar network

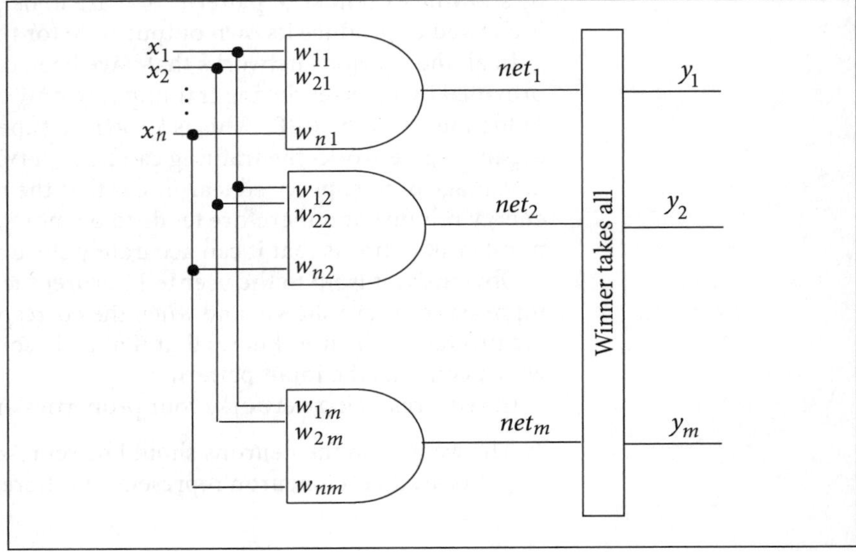

The weighted sum can be expressed in vector terms as:

$$net_j = \sum_{i=0}^{n} w_{ij}x_i = [\mathrm{X}][\mathrm{W}_j] = |\mathrm{X}||\mathrm{W}_j|\cos(\theta)$$

where |X| means the magnitude of [X]. In other words, net_j is the product of two vectors and can therefore be expressed as the 'length' of one vector multiplied by the projection of the other vector along the direction of the first vector. To illustrate, try a two-input problem, where $x_1 = 1$, $x_2 = 0.3$, and $w_1 = 1$, $w_2 = 1$.

$$[x_1 \quad x_2]\begin{bmatrix} w_1 \\ w_2 \end{bmatrix} = w_1x_1 + w_2x_2$$

$$[1 \quad 0.3]\begin{bmatrix} 1 \\ 1 \end{bmatrix} = 1 + 0.3 = 1.3$$

$$|\mathrm{X}| = \sqrt{x_1^2 + x_2^2} = \sqrt{1^2 + 03^2} = 1.044$$

$$|\mathrm{W}| = \sqrt{w_1^2 + w_2^2} = \sqrt{1^2 + 1^2} = 1.414$$

The vectors [X] and [W] can be represented in a diagram, as shown in Figure 7.2. Their lengths are 1.044 and 1.414 respectively, and the angle between them, θ, is about 28°. Therefore, $\cos(\theta) = 0.883$, and the product of the two vectors is:

$$|\mathrm{X}||\mathrm{W}|\cos(\theta) = 1.303$$

The figures for the weighted sum and for the vector product come out more or less the same, namely 1.3.

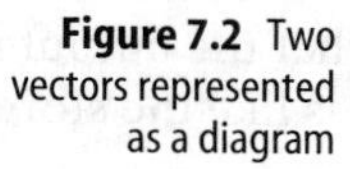

Figure 7.2 Two vectors represented as a diagram

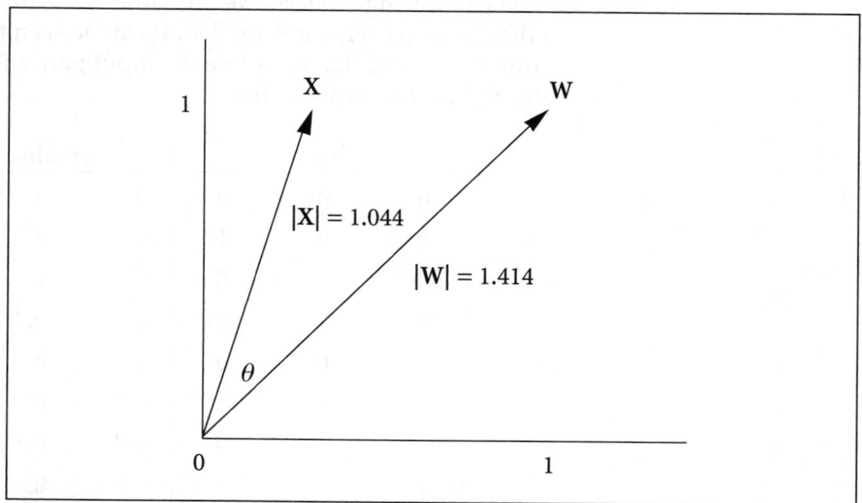

If the two vectors, [X] and [W_j], are normalized, which means scaling them so that they each have a length of 1, the product ends up being equal to cos(θ), where θ is the angle between the two vectors.

If the two vectors are identical, θ will be 0, and cos(θ) = 1. The further apart the two vectors become, the greater the angle (positive or negative) and the smaller the value of the product. In the extreme, where the input pattern is the inverse of the stored weights, θ is ±180° and cos(θ) = –1. If the assumption is made that patterns that are similar will be close together in pattern space, then normalizing the input vector means that the output of a neuron is a measure of the similarity of the input pattern and its weights.

If a network is set up initially with random weights, when an input pattern is applied, each of the neurons will produce an output which is a measure of the similarity between the weights and the input pattern. The neuron with the largest response will be the one with the weights that are most similar to the input pattern.

Normalising the input vector means dividing by the magnitude of the vector which is the square root of the sum of the square of all the elements in the vector.

$$|X| = \sqrt{\Sigma_{i=1}^{n} x_i^2}$$

As an example, assume that a neuron has been trained with the pattern:

0 1 1

and that the weights are normalized. The magnitude of [X] is:

$$|X| = \sqrt{\Sigma_{i=1}^{n} x_i^2} = \sqrt{0^2 + 1^2 + 1^2} = \sqrt{2} = 1.412$$

The weights are therefore:

$$w_1 = 0 \qquad w_2 = w_3 = \frac{1}{1.412} \approx 0.7$$

The following table shows the value of the output of this neuron when other input patterns are applied. It can be seen that the output ranges from 0 to 1, and that the more the input pattern is like the stored pattern the higher the output score.

Input			Normalised input			Output
0	0	0	0	0	0	0
0	0	1	0	0	1	0.7
0	1	0	0	1	0	0.7
0	1	1	0	0.7	0.7	1
1	0	0	1	0	0	0
1	0	1	0.7	0	0.7	0.5
1	1	0	0.7	0.7	0	0.5
1	1	1	0.6	0.6	0.6	0.8

The next step is to apply a learning rule so that the neuron with the largest response is selected and its weights are adjusted to increase its response. The first part is described as a 'winner takes all' mechanism, and can be simply stated as:

$$y_j = 1 \qquad \text{if } net_j > net_i \text{ for all } i, i \neq j$$

$$y_j = 0 \qquad \text{otherwise}$$

In Chapter 5, the Hamming network was shown, which could implement this function. There are a number of alternative ways which will not be discussed in this book but can be found elsewhere (Grossberg, 1973).

The learning rule for adjusting the weights is different from the Hebbian rules that have been described up to now. Instead of the weights being adjusted so that the actual output matches some desired output, the weights are adjusted so that they become more like the incoming patterns.

Mathematically, the learning rule, which is often referred to as Kohonen learning (Hecht-Nielsen, 1987), is:

$$\Delta w_{ij} = k(x_i - w_{ij})y_j$$

In the extreme case where $k = 1$, after being presented with a pattern the weights in a particular neuron will be adjusted so that they are identical to the inputs, that is $w_{ij} = x_i$. Then, for that neuron, the output is maximum for that input pattern. Other neurons are trained to be maximum for other input patterns.

Alternatively, with $k < 1$ the weights change in a way that makes them more like the input patterns, but not necessarily identical. After training the weights should be representative of the distribution of the input patterns.

Notice that the term y_j is included so that, during training, only the neuron with the largest response will have an output of 1 after competitive learning, while all other outputs are set to 0. Therefore only this neuron will adapt its weights.

The combination of finding the weighted sum of normalized vectors, Kohonen learning and competition means that the instar network has the ability to organise itself such that individual neurons have weights that represent particular patterns or classes of patterns. When a pattern is presented at its input, a single neuron, which has weights that are the closest to the input pattern, produces a 1 output while all the other neurons produce a 0. Learning in the instar is therefore unsupervised.

The outstar network is shown in Figure 7.3, and again can be seen to consist of neurons which find a weighted sum of the inputs. Its function is to convert the input pattern, x_i, into a recognised output pattern and is therefore supervised. It is assumed that the input pattern consists of a single 1, with all other inputs set to 0, just like the output of the instar. Consequently, the weighted sum in each neuron, net_j, is equal to w_{ij} when x_i is the input that is set to 1. For example, if x_2 is the input that is 1, and all other inputs are 0, the outputs will be:

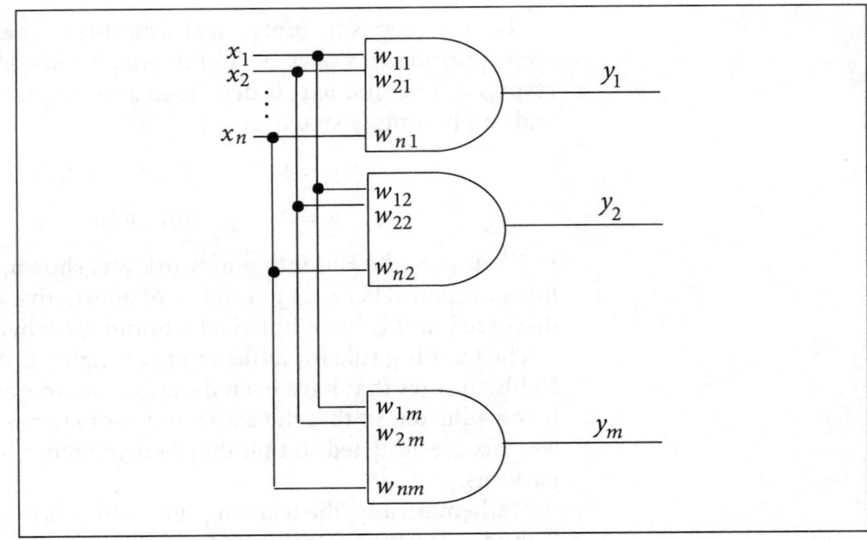

Figure 7.3 The outstar network

$$y_1 = w_{21}$$

$$y_2 = w_{22}$$

$$y_3 = w_{23}$$

If the values of the weights are again made equal to a pattern consisting of 0s and 1s, the output pattern will be that stored pattern. So, the 1 on the input recalls a previously stored pattern.

The learning rule for the outstar network is often referred to as Grossberg learning (Hecht-Nielsen, 1987) and can be stated mathematically as:

$$\Delta w_{ij} = k(y_i - w_{ij})x_i$$

This rule is complementary to the instar rule in that the weights are now adjusted so that they will eventually equal the desired output value, and that only the weights associated with the input that is a 1 are adjusted.

This learning rule is in effect the delta rule, described in Chapter 2, if it is remembered that only one input is 1 while all others are 0. Then w_{ij} is equivalent to $\Sigma w_{ij} x_i$.

Since the outstar network only works if one of its inputs is 1 and all others are 0 it is possible to join the instar and outstar networks together, as shown in Figure 7.4.

The network in Figure 7.4 has been trained and has therefore stored the patterns 1 0 0, 1 1 0 and 0 0 1, with $k = 1$ and the weights normalized. For example, the weights in neuron 2 are 0.7, 0.7 and 0.

If the pattern 1 0 0 is presented to the network, neuron 1 in the first layer will produce the maximum value output of 1 while the others produce 0.7 and 0 respectively. The network is allowed to settle so that the output of neuron 1 becomes 1 while the other outputs become 0. The inputs to the second layer of neurons will then be 1 0 0, giving an output of 1 0 0.

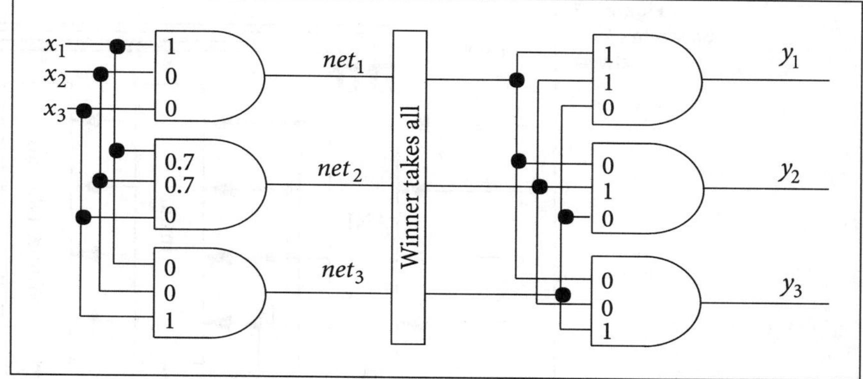

Figure 7.4 Instar/outstar network

A property of this network is that if a new pattern is presented, the stored pattern that is most similar to it will produce the maximum output in the first layer and then recall the stored pattern in the second layer. So the instar/outstar network can generalise and recall perfect data from imperfect data.

In this form, the network is equivalent to the forward-only counter-propagation network (Hecht-Nielsen, 1987). The code stored in the outstar network can be selected by the user, so this form of network is supervised. The following section goes on to discuss Grossberg's ART1 network, which is unsupervised.

7.3 Adaptive resonance theorem (ART)

The ART1 network (Grossberg, 1976) is a good example of a self-organising network. It is similar to the forward-only counter-propagation network except that it has some extra features. It uses some extra 'circuitry', as shown in Figure 7.5, to perform competitive learning and 'vigilance', ρ, which will be explained later.

When the network is operating, it behaves as a 'follow my leader' classifier (Lippmann, 1987). It receives input patterns and produces at its output the classification code for those patterns. There are no distinct phases of learning and then operating: it does both at the same time.

The way that ART1 works can be described by the following steps:

Step 1 Input pattern X directly to the instar network.

Step 2 Find the neuron with the maximum response – neuron i.

Step 3 Make the output of neuron i equal to 1 and all others 0.

Step 4 Feed the output of the instar to the input of the outstar to generate an output pattern, Y.

Step 5 Feed Y back to create a new pattern which equals X AND Y.

Step 6 Calculate the vigilance, ρ.

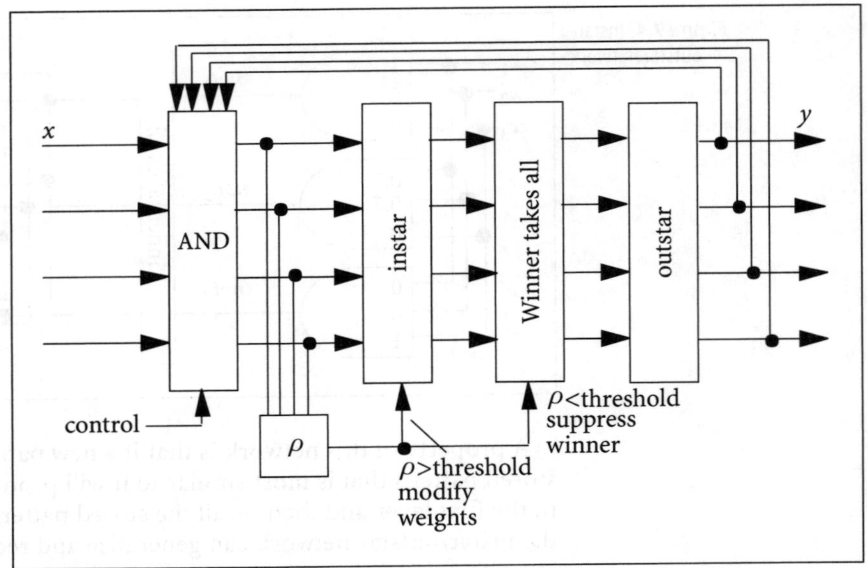

Figure 7.5 Grossberg's ART1 network

Step 7 If ρ is greater than some predetermined threshold, modify the weights of neuron i in the instar network so that they are normalized versions of the new pattern X AND Y. Also, in the outstar network, modify the weights so that the output produced equals the new pattern X AND Y. Go back to step 1.

Step 8 If ρ is less than the threshold, suppress the output of neuron i, and find the neuron with the next largest output value – neuron j. Go to step 3.

All neurons in the instar start off with some arbitrary weights on their inputs. It has been suggested (Lippmann, 1987) that if the neurons in the first layer are initialized with weights of $1/(n + 1)$, then, no matter what patterns appear at the input, the response of at least one of the neurons which has been previously adjusted will be larger. The connections to the neurons in the output layer should all be set to 1. This is so that when a new class of pattern is introduced to the network, the initial output pattern will be all 1s and the AND of the output pattern with the input pattern will equal the input pattern, so this will get stored.

When the first pattern arrives, the neuron that produces the largest response is selected using a winner takes all mechanism to ensure that this neuron alone in the first layer has its weights adjusted. All of the neurons in the second layer will have the weight associated with the input connected to this single neuron adjusted to produce the same pattern at the output of the network. Thus the first pattern that the network receives is regarded as the template or exemplar pattern for the first class.

When subsequent patterns arrive at the input, the neuron that produces the largest response is selected by the winner takes all mechanism; then it will do either of two things:

1 If the pattern is similar to the exemplar pattern (measured by the vigilance, ρ), a new exemplar is produced which is a combination of the old exemplar and the new input pattern.

2 If the pattern is dissimilar by the same measure, ρ, to the exemplar pattern, the new pattern becomes the exemplar for a new class.

This continues forever, with new classes being added when necessary, and the existing exemplars being modified so that they become more representative of the class that they exemplify.

The outputs are the exemplars themselves. So, at any stage in the operation of the network, an input pattern will produce an output pattern which is the exemplar for the class in which the input pattern belongs.

Let us first look at the situation in which a new input pattern is presented at the input which is similar enough to one of the stored patterns to be regarded as belonging to the same class, but which is not identical to it. To start with, the new pattern is input to the instar network directly, to produce the maximum response in one of the neurons. This generates a stored pattern, Y, at the output of the outstar network. The AND of the input pattern, X, and the stored pattern, Y, is found. At this point the vigilance, ρ, is measured to see if it is above or below some pre-set value or threshold. The vigilance equals the number of 1s in the pattern produced by finding X AND Y, divided by the number of 1s that are in the input pattern, X. This can be written as:

$$\rho = \frac{\Sigma_{i=1}^{n} x_i \wedge y_i}{\Sigma_{i=1}^{n} x_i}$$

where y_i is the stored pattern in 0/1 notation and $\wedge$ is the AND function. When there is a perfect match, the value of ρ is 1, otherwise it is between 0 and 1. If the vigilance is above the threshold, the adapted pattern is stored in the network.

When this happens the neuron in the first layer that has been selected has its weights adjusted so that they match the AND of the input pattern and the old exemplar pattern for this class and are then normalized. In ART1, normalisation means dividing the weights by the sum of the value of the elements in the vector rather than the sum of the squares. The weights are therefore given as:

$$w_i = \frac{L(x_i \wedge y_i)}{L - 1 + \Sigma_{j=1}^{n}(x_j \wedge y_j)}$$

where L must be greater than 1 (Carpenter and Grossberg, 1987a). A typical solution is to let $L = 2$ so that the equation becomes:

$$w_i = \frac{2(x_i \wedge y_i)}{1 + \Sigma_{j=1}^{n}(x_j \wedge y_j)}$$

In the second layer, weights w_{ij} in each of the neurons are adjusted so that they too correspond to the AND of the two patterns and therefore have values of either 0 or 1. The effect is that the patterns 'resonate', producing a stable output.

Take the network that was shown in Figure 7.4 again. If the network has learned the three patterns 1 0 0, 1 1 0 and 0 0 1, the weights in the three neurons (from top to bottom) after normalisation are 1 0 0, 0.67 0.67 0 and 0 0 1.

Assume that an input pattern of 0 1 0 is presented to the network. The response of the three neurons is 0, 0.67 and 0 respectively; therefore the second neuron gives the largest response, indicating that its stored pattern is the closest to the input pattern. The winner takes all network produces a 1 at the outputs of neuron 2, and 0 at each of the outputs of neuron 1 and 3. Thus, the pattern 1 1 0 is produced at the output of the ART network. This is fed back, and the AND of the two patterns calculated:

Input pattern X	0	1	0
Output pattern Y	1	1	0
$X \wedge Y$	0	1	0

The number of 1s in X is 1, and the number of 1s in $X \wedge Y$ is also 1, so the vigilance for this match is 1/1 = 1. Assuming a typical value for the threshold of say 0.7, then as the vigilance is larger than this, the pattern is accepted into the class and the new exemplar pattern stored.

The weights in neuron 2 are modified so that they now represent the AND of the two patterns after normalisation, as shown in Figure 7.6, where the extra vigilance circuitry has been removed for clarity.

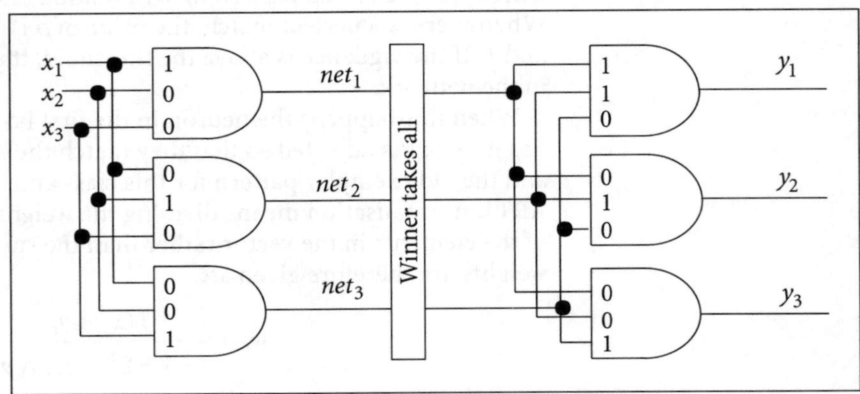

Figure 7.6 Weight modification in the ART1 network

The weights in the output layer are also modified so that if neuron 2 produces a 1 output then the exemplar pattern will be produced at the output of the network.

Now let us look at what happens when the vigilance falls below the threshold. The ART1 network has the ability to create a new class by adding a new neuron to the input layer which is fully connected to the output layer. It is simpler to assume that this neuron already exists. When a new pattern arrives at the inputs, the neurons in the first layer respond and the one with the highest response is chosen. If the value of the vigilance is below the threshold, the neuron that is responding is suppressed and the neuron with the next highest response is selected. This continues until all of the previously adjusted neurons have been tried, at which point one of the unused input neurons is selected and its weights set to maximise its response to this new pattern. The weights in the output layer are also modified so that the pattern will be reproduced at the output.

Figure 7.7 shows the network of Figure 7.4 again, with one more neuron shown in the input layer and additional weighted connections to the output layer. When the input pattern 0 1 1 is applied, neuron 3 gives the highest response of 1. The vigilance is only 0.5, so, assuming that this is below the threshold, a new class has to be created. This is done by 'turning on' one of the unused neurons. After normalisation the weights in the new neuron are 0, 0.67 and 0.67. The weights in the output layer are also modified so that the new pattern will be output when appropriate.

Figure 7.7 Addition of a new class

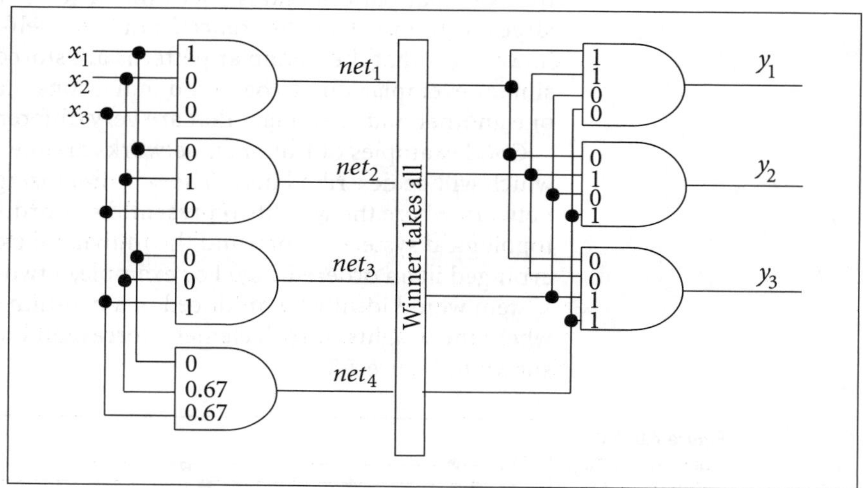

ART1 has only briefly been described here. There is a lot more to it than has been suggested. For example, the features of the ART networks, such as the vigilance measure, ρ, the winner takes all mechanism, the AND function of the input and output patterns and finally the suppression of the output if a new class is created are all relatively simple to describe and to simulate. However, Grossberg has found ways to implement these features using neural models rather than computer algorithms. We will not go into any more detail about how these are implemented, but anyone interested in Grossberg's work should take time to study it, much of which can be found in his books (Grossberg, 1987, 1988).

The ART1 network has been extended to ART2 (Carpenter and Grossberg, 1987b) and recently ART3 (Carpenter and Grossberg, 1990). These allow more complex input patterns, such as analog patterns, to be processed. However, there is insufficient space in this book to give details of these networks.

7.4 Kohonen networks

Teuvo Kohonen has produced many research papers and books on the subject of associative memory and self-organising systems (Kohonen, 1984). He has provided very rigorous mathematical models for these systems which, as he himself has said, have often been 'rediscovered' (Kohonen, 1988). However, the networks which now tend to be called Kohonen networks are only a small area of his research.

The aim of a Kohonen network is to produce a pattern classifier, as before, which is self-organising, using Kohonen learning to adjust the weights. Typically, a Kohonen network consists of a two-dimensional array of neurons with all of the inputs arriving at all of the neurons. Each neuron has its own set of weights which can be regarded as an exemplar pattern. When an input pattern arrives at the network, the neuron with the exemplar pattern that is most similar to the input pattern will give the largest response. One difference from other self-organising systems, however, is that the exemplar patterns are stored in such a way that similar exemplars are to be found in neurons that are physically close to one another and exemplars that are very different are situated far apart.

Good examples of Kohonen networks are the self-organising maps which will be described here. The self-organising maps aim to produce a network where the weights represent the coordinates of some kind of topological system or map and the individual elements in the network are arranged in an ordered way. For example, a two-dimensional coordinate system would ideally be produced in a two-dimensional array of elements where the weights in each element correspond to the coordinates, as shown in Figure 7.8.

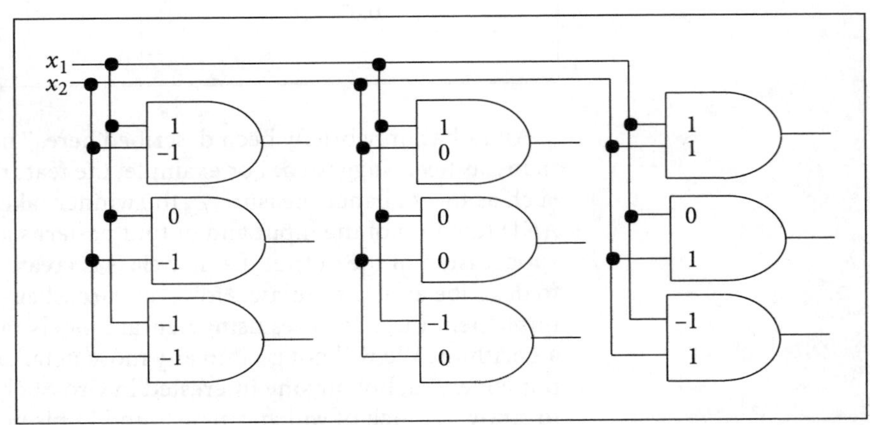

Figure 7.8 A two-dimensional map represented as a two-dimensional array of neurons

Initially, each weight is set to some random number. Then pairs of randomly selected coordinates are presented to the system with no indication that these coordinates are taken from a square grid. The system then has to order itself so that the weights correspond to the coordinates, and so that the position of the element also corresponds to the position in the coordinate system.

The method for achieving this is to use a matching procedure like the one used in the instar network. Various matching criteria can be used, but one that is often used is the Euclidean distance. This is found by taking the square root of the sum of the squared differences.

$$D_j = \sqrt{\Sigma_{i=1}^{n}(x_i - w_{ij})^2}$$

For a two-dimensional problem, the distance calculated in each neuron is:

$$D_j = \sqrt{\Sigma_{i=1}^{2}(x_i - w_{ij})^2} = \sqrt{(x_1 - w_{1j})^2 + (x_2 - w_{2j})^2}$$

An input vector is simultaneously compared to all of the elements in the network, and the one with the lowest value for D is selected. If this element is denoted with a c, then a neighbourhood around c is also defined as being those elements which lie within a distance of N_c from c. The exact nature of the neighbourhood can vary, but one that is frequently used is shown in Figure 7.9.

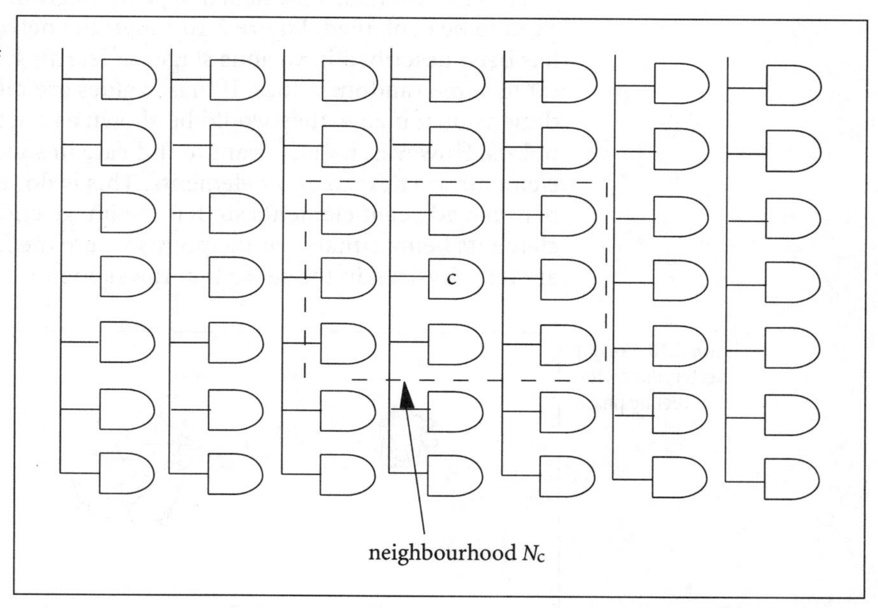

Figure 7.9 The neighbourhood of element c

Having identified the element c, the centre of the neighbourhood and the elements that are included in the neighbourhood, the weights of those elements are adjusted using Kohonen learning, as described in Section 7.1. The weights of all other elements are left alone. The formula for weight adjustment using Kohonen learning is repeated here:

$$\Delta w_{ij} = k(x_i - w_{ij})y_j$$

The decisions about the size of N_c and the value of k are important. First of all, both must decrease with time, and there are several ways of doing this. It has been shown that there is a strong relationship between Kohonen learning and Kalman filters, particularly in the way that both use a constant, k, that decreases from 1 to 0 over time (Picton, 1991b).

The value of k and the size of N_c could decrease linearly with time; however, it has been pointed out that there are two distinct phases – an initial ordering phase, in which the elements find their correct topological order, and a final convergence phase in which the accuracy of the weights improves. For example (Kohonen, 1988), the initial ordering phase might take 1000 iterations where k decreases linearly from 0.9 to 0.01, say, and N_c decreases linearly from half the diameter of the network to one spacing. During the final convergence phase k may decrease from 0.01 to 0 while N_c stays at one spacing. This final stage could take from 10 to 100 times longer than the initial stage, depending on the desired accuracy (Kohonen, 1984).

Let us return to the example that was described earlier, where a two-dimensional array of elements was arranged in a square. In that example the aim was to map a rectangular two-dimensional coordinate space onto this array, which is the simplest case to imagine. The network is capable of more complex mappings.

To illustrate this, a particular type of diagram will be used, which will need to be explained. Figure 7.10 shows the network for the example that has been described in various stages of learning. Initially, the weights are set to some random values. If these values are plotted on a two-dimensional image, they would be shown as a set of randomly distributed points. However, we also want to indicate in some way, that some elements are next to other elements. This is done by drawing a line between adjacent elements so that the image ends up as a set of lines, the elements being situated at the points where the lines intersect. These lines are not physical, in the sense that the elements are not joined together.

Figure 7.10 Weight vectors during the ordering phase

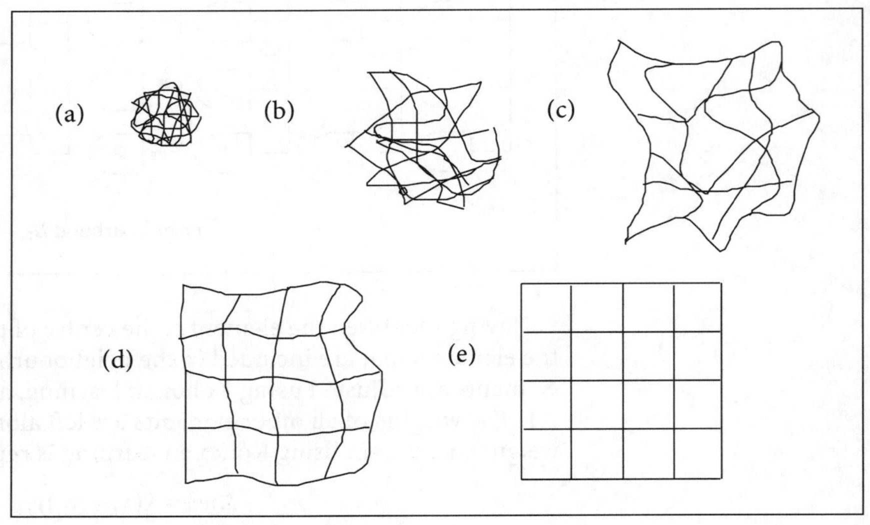

The system is presented with a set of randomly chosen coordinates. As time elapses, the weights order themselves so that they correspond to the positions in the coordinate system. Another way of thinking about this is that the weights distribute themselves in an even manner across the coordinate space so that, in effect, they learn to 'fill the space'.

This, and other examples (Kohonen, 1984), show how two-dimensional arrays which map onto a coordinate system can arrange the weights so that the 'nodes' in that system are distributed evenly. One thing that has not been mentioned yet is the output. What is the output of a Kohonen network?

At the start of this book it was said that neural networks are pattern classifiers. Even this self-organising map can be a pattern classifier if the weights are considered as the coordinates in pattern space. Training involves grouping similar patterns in close proximity in this pattern space, so that clusters of similar patterns cause neurons to fire that are physically located close together in the network.

In the simplest case, the output is the weighted sum of the inputs, possibly after passing through a sigmoid function. If this is the case, when an input is applied, the output will be a single 1 in the region of the coordinate space which corresponds to a particular class of patterns. Clearly, the outputs need to be interpreted, but it should be possible to identify which regions belong to which class by showing the network known patterns and seeing which areas are active.

The output need not be two-dimensional, even though the layout of the physical devices might be two-dimensional. If there are n weights, then each weight corresponds to a coordinate. So although a two-dimensional image of which elements are active is produced when patterns are presented to the input, the interpretation of that map might have many more dimensions.

For example, a system where each element has three weights would organise itself so that the different pattern classes occupy different parts of a three-dimensional space. If the network is observed, only individual elements firing would be seen, so it is misleading to think in terms of the physical layout.

7.5 Neocognitron

The neocognitron (Fukushima, 1980), and its predecessor the cognitron (Fukushima, 1975), are devices that were developed for the specific task of recognising handwritten characters. It is tolerant of distortions and shifts in position of the characters. Figure 7.11 shows a schematic diagram of the network.

The neocognitron is a multi-layered network which consists of an input layer which holds a binary image, followed by alternating S and C layers.

The cells in the S layers are shape detectors which pick out specific features. The ones near the input tend to pick out low-level features, such as lines and angles, whereas those in higher levels tend to pick out whole patterns. The cells behave more or less like the conventional McCulloch–

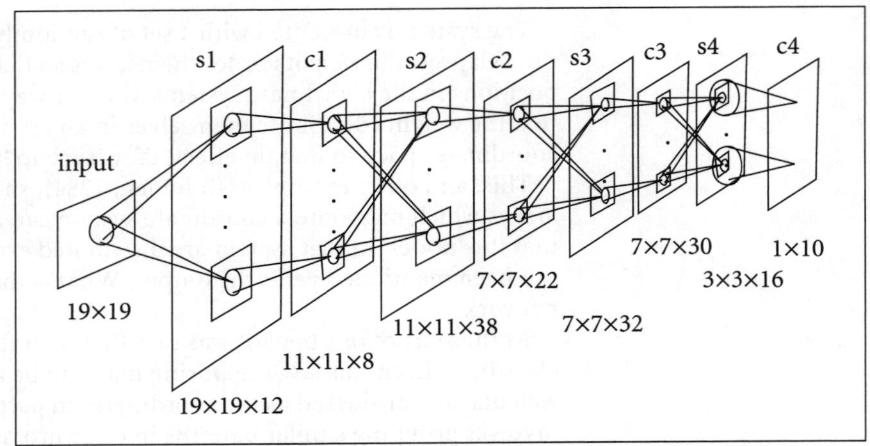

Figure 7.11 Neocognitron

Pitts model of a neuron, as described in Chapter 1. Figure 7.12 shows the detail of an S cell which is connected to three cells in the input layer.

Inputs are weighted and summed, and then passed through a non-linear output function. Most of the inputs are excitory, but one input, v, is always inhibitory and its weighted value, h, is always subtracted from the weighted sum.

An additional feature is that the sum is divided by $1 + h$, so that the final output, if positive, is scaled by the amount of inhibition that is present. The weights in the S cells are variable and are adjusted during the learning phase. Note that the weight on the inhibitory input is also variable.

The C cells are different in that their weights are fixed. The C-layers are simply there to provide invariance to shift. Each cell in the C-layer responds to a set of S cells from the previous layers which detect the same pattern. A C cell becomes active if at least one of the S cells is active. Thus, no matter where the pattern appears in a region of the image, at least one of the S cells should detect it, and consequently the C cell becomes active.

The learning rule for the adjustable weights in the S cells is winner takes all again and is described as follows: a weight is increased if the output of the S cell is the maximum in its vicinity and the input associated with that weight is active.

Training can be supervised or unsupervised. If it is unsupervised, the cells with the maximum output response are always chosen as the 'seed'

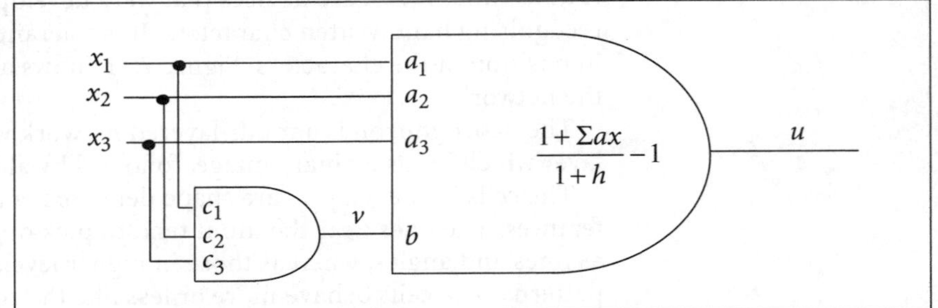

Figure 7.12 An S cell

for the next layer. Supervised training is useful when the aim is to classify patterns that are not necessarily similar but which are deemed to be the same by convention or style. For example, upper- and lower-case letters might be regarded as belonging to the same class, even though the characters are often quite different. During supervised training, the 'teacher' selects the seed for the next layer.

The detailed operation of the neocognitron is quite complex, but it is worth looking at the operation of the cells in the S-layer in more detail (Fukushima *et al.*, 1983).

The output, u, is given by the equation:

$$u = r\phi\left(\frac{1+\Sigma_{i=1}^{3} a_i x_i}{1+[rbv/(1+r)]} - 1\right)$$

$$\text{where } \phi(x) = x \text{ when } x \geq 0$$

$$\text{and } \phi(x) = 0 \text{ when } x < 0$$

$$v = \sqrt{\Sigma_{i=1}^{3} c_i x_i^2}$$

and the values of c_i are chosen such that they decrease monotonically as their distance from the S cell increases and are normalized such that the following equation is satisfied:

$$\sum_{i=1}^{3} c_i = 1$$

A quantity, s, is defined as:

$$s = \frac{\Sigma_{i=1}^{3} a_i x_i}{bv}$$

Then, when all the values of a_i and b are large, the equation for u reduces to:

$$u \approx r\phi\left(\frac{(r+1)s}{r} - 1\right)$$

If this S cell has been trained on pattern P1, when it is shown another pattern, P2, the value of s can be interpreted as the scalar product of the normalized vectors P1 and P2 (Fukushima, 1989). In other words, s gives the cosine of the angle between the two vectors and therefore has a value between 0 and 1.

When the patterns are identical, $s = 1$, and the output of the cell is also 1. When the patterns are similar, the value of s is close to 1. The cell fires if the value of:

$$\frac{(r+1)s}{r} - 1 \geq 0$$

$$\text{or } s \geq \frac{r}{r+1}$$

When r is large, the value of $r/(r+1)$ is close to 1, so the cell only fires when the patterns are very similar. When r is smaller the cell fires for a wider range of patterns. Therefore r alters the selectivity of the cell.

The weights of the cell that gives the maximum response are adjusted as:

$$\Delta a_i = qc_i x_i$$

$$\Delta b = qv$$

where q is a positive constant. All adjustable weights are initially set to 0.

As an example, let $a_1 = 0$, $a_2 = 0$, $b = 0$, $c_1 = c_2 = 0.5$, $r = 1$, $q = 0.5$. When the pattern $x_1 = 0$, $x_2 = 1$ is shown, the response is:

$$v = \sqrt{c_1 x_1^2 + c_2 x_2^2} = 0.7$$

$$u = r\phi\left(\frac{1 + a_1 x_1 + a_2 x_2}{1 + [rbv/(r+1)]} - 1\right) = 0$$

$$\Delta a_1 = qc_1 x_1 = 0$$

$$\Delta a_2 = qc_2 x_2 = 0.25$$

$$\Delta b = qv = 0.35$$

So $a_1 = 0$, $a_2 = 0.25$ and $b = 0.35$. If these values are substituted, then the new value of u is 0.114. Continuing like this, the values of a_2 and b get progressively larger and the value of u approaches 1.

Initially, an S-cell is chosen arbitrarily to be the pattern detector. When its weights are adjusted it becomes the cell that produces the maximum response, and so adjusts its weights to reinforce the response. In this way, the various cells become tuned to small primitive features.

CHAPTER SUMMARY

This chapter has discussed a number of self-organising systems. In each case there were neurons that find the weighted sum of the inputs, and this sum is used as a matching score between the neurons and the input pattern. The neuron with the largest response becomes the only active neuron by the application of some winner takes all mechanism, and its weights are adjusted so that its response is increased.

Variations include the ART1 network, which uses vigilance to determine whether a new input pattern should be absorbed into an existing class or used as the exemplar for a new class. The Kohonen network attempts to physically arrange its neurons so that the patterns that it stores are arranged such that similar patterns are close to each other and dissimilar patterns are far apart. Finally, the neocognitron selects its own feature primitives and classifies them, but then also allows detection of similar patterns even when they are shifted.

The instar and outstar networks are used in the counter-propagation network, which is sometimes used in applications such as handwriting interpretation. The ART networks are less popular in applications, probably because they were developed as research into cognitive models of biological functions such as memory, so applications are really a spin-off. They do have some useful features, however, such as their ability to learn very quickly. The neocognitron was a one-off design, built to read handwritten characters, and has not been used for other applications. But by far the most popular of all the networks described in this chapter is the Kohonen network. The description in this chapter showed the network's ability to self-organise and to classify. This has led to the Kohonen network being used in diverse applications, including speech recognition, image analysis and control (Ritter *et al.*, 1992). This latter application will be discussed in the next chapter.

SELF-TEST QUESTIONS

1. For the network shown in Figure 7.13, what will the response be to the input values shown?
2. An ART network has learned the following three patterns:

   1000, 1111, 0010

   (a) What are the values of the normalized weights in the three neurons that store these patterns in the instar layer, assuming that $L = 2$?

   (b) What are the weights in the outstar layer?

   (c) What would the initial output be if the ART network is presented with the pattern 1110?

   (d) What would the value of the vigilance be, and what is likely to happen next?
3. In Kohonen learning and in Grossberg learning, what happens to the weights if the constant, k, is set to 1?
4. The input pattern [X] = [1 0 1] and a set of weights [W] = [0.2 1.5 2.0] in a neuron can be interpreted as two vectors.

   (a) What are the lengths of the two vectors and what is the value of the angle between them?

   (b) What is the Euclidean distance between them?

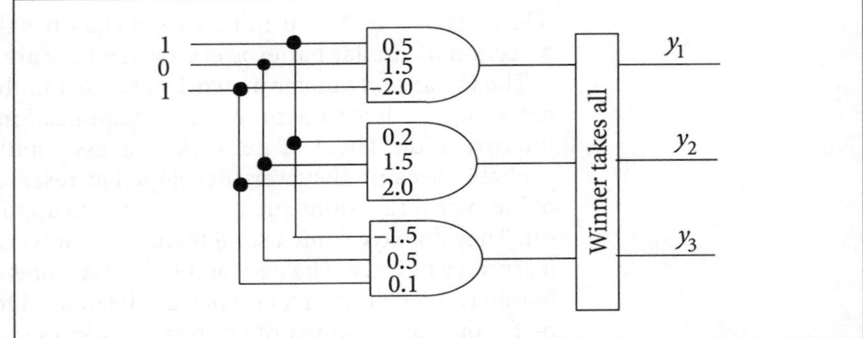

Figure 7.13 Network for Question 1

SELF-TEST ANSWERS

1 With these input values, the outputs of the three neurons are, from the top:

$$0.5 - 2.0 = -1.5$$

$$0.2 + 2.0 = 2.2$$

$$-1.5 + 0.1 = -1.4$$

Therefore the middle neuron wins the competition, so that the values for y are:

$$y_1 = 0, y_2 = 1, y_3 = 0$$

2 (a) Applying the following equation:

$$w_i = \frac{2(x_i \wedge y_i)}{1 + \Sigma_{j=1}^{n}(x_j \wedge y_j)}$$

the normalized weights are 1000 in the first neuron, 0.4 0.4 0.4 0.4 in the second and 0010 in the third.

(b) The outstar layer would have four neurons in order to produce 4-bit output patterns. The weights would be 110, 010, 011 and 010.

(c) The pattern 1110 would produce the values 1, 1.5 and 1 in the three neurons in the instar layer. The winner takes all mechanism would select neuron 2 as the closets and would therefore give it an output of 1 and all other neurons an output of 0. When this is fed to the outstar layer it produces an output of 1111, which represents the nearest stored pattern

(d) The pattern 1111 is fed back to the input, where it is compared with the input and the AND of the two patterns found. In this case the AND of 1110 and 1111 is 1110. The vigilance is found using the equation:

$$\rho = \frac{\Sigma_{i=1}^{n} x_i \wedge y_i}{\Sigma_{i=1}^{n} x_i}$$

Since the input pattern and the AND of the output and the input pattern are the same, they contain the same number of 1s, so the top and bottom of the equation are the same, giving a vigilance of 1. Since the threshold is usually less than 1, a vigilance of 1 will almost certainly result in the AND of the patterns being stored. The second neuron therefore has its weights modified so that it contains the normalized version of the AND of the input and output. The weights become 0.5 0.5 0.5 0, and all of the neurons in the outstar are modified to 110, 010, 011 and 000.

3 The equation for Kohonen learning is:

$$\Delta w = k(x - w)y$$

The weights are updated, so:

$$w(\text{new}) = w + \Delta w$$

Only the weights in the winning neuron are updated, so for the winning neuron, $y = 1$. If $k = 1$, the equation becomes:

$$\Delta w = (x - w)$$

$$\text{and } w(\text{new}) = w + (x - w) = x$$

This means that for the winning neuron, the weights take on the values of the input pattern.

Similarly, for Grossberg learning, the equation is:

$$\Delta w = k(y - w)x$$

Only the weights associated with an input value of 1 are adjusted, so that when $k = 1$, the value of the weights are set to equal the desired output values.

4 (a) The length of a vector is found by squaring all of the elements of the vector, summing the squares then finding the square root.

$$\text{Length of [X]} = |X| = \sqrt{1^2 + 0^2 + 1^2} = \sqrt{2} = 1.414$$

$$|W| = \sqrt{0.2^2 + 1.5^2 + 2.0^2} = \sqrt{6.29} = 2.508$$

$$\text{Product of the two vectors} = [X][W]^t = [1 \quad 0 \quad 1]\begin{bmatrix} 0.2 \\ 1.5 \\ 2.0 \end{bmatrix} = 2.2$$

Since the product is also $|X||W|\cos\theta$, then:

$$2.2 = |X||W|\cos\theta = 1.414 \times 2.508 \times \cos\theta$$

$$\cos\theta = 2.2/(1.414 \times 2.508) = 0.62$$

Therefore $\theta = 51.657°$

(b) The Euclidean distance, D_j, for neuron j is given as:

$$D_j = \sqrt{\Sigma_{i=1}^{n} (x_i - w_{ij})^2}$$

The input pattern [X] = [1 0 1] and the weights are [W] = [0.2 1.5 2.0].

$$D_j = \sqrt{(1-0.2)^2 + (0-1.5)^2 + (1-2.0)^2} = 1.97$$

CHAPTER 8

Neural networks in control engineering

CHAPTER OVERVIEW

This chapter gives a description of how neural networks are used to control complex systems. It explains:

- how neural networks can learn to control systems by punishing or rewarding performance
- how reinforcement learning works
- how a recurrent network can model the dynamics of a system.

8.1 Introduction

In the design of classical control systems there are three distinct steps:

Step 1 Find a mathematical model of the system that you wish to control.

Step 2 If the model is linear then move on to step 3, otherwise try to alter the system or its model so that it becomes linear.

Step 3 Apply some technique from control theory to produce the desired system responses.

The second step of this procedure is clearly the problematic one. It is not always possible to get a linear model, and non-linear systems are not as well understood as linear systems. Methods exist only for the design of some of the simpler non-linear systems.

Neural networks provide a useful way of controlling complex non-linear systems because a model of the system is not required. One way of achieving this when a model is not available is to use failure avoidance. An early example of this is called Michie's boxes. It is a method of controlling a non-linear system without a model and using some form of weight adjustment mechanism. Although this is not a neural network like those that have been described so far, it does share some common features.

8.2 Michie's boxes

Michie's boxes is a method that was developed by Donald Michie and his colleagues at Edinburgh University (Michie and Chambers, 1968). It was applied to a very difficult control engineering problem, namely balancing a pole on a cart.

The pole is hinged to the cart, but it and the cart can only move in two dimensions. In terms of classical control engineering, this is a difficult problem because, unless certain assumptions are made (namely that the mass of the cart is much greater than the mass of the pole and that the angular displacement of the pole is small), the cart and pole can only be modelled with non-linear differential equations.

Figure 8.1 shows the problem, which can be modelled using a set of differential equations (Barto *et al.*, 1983). The dot notation is used as shorthand to mean the derivative with respect to time.

Figure 8.1 The cart and pole

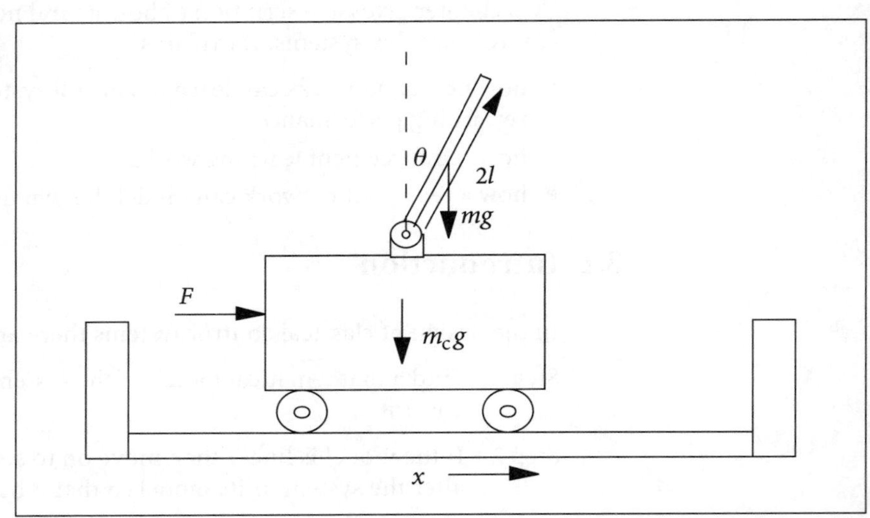

$$\ddot{\theta} = \frac{g\sin\theta + \cos\theta[-F - ml\dot{\theta}^2\sin\theta + \mu_c\,\mathrm{sgn}(\dot{x})] - \mu_p\dot{\theta}/ml}{l[(4/3) - m\cos^2\theta/(m_c + m)]}$$

$$\ddot{x} = \frac{F + ml[\dot{\theta}^2\sin\theta - \ddot{\theta}\cos\theta] - \mu_c\,\mathrm{sgn}(\dot{x})}{m_c + m}$$

where
g = acceleration due to gravity
m_c = mass of cart
m = mass of pole
l = half-pole length
μ_c = coefficient of friction of cart on track
μ_p = coefficient of friction of pole on cart
F = force applied to cart's centre of mass

To balance the pole, a force of fixed value F can be applied in either direction at discrete moments in time. In addition, the cart must stay within the limits of the track, so it cannot just move forever in one direction or the other.

There are four variables in this problem, which are quantized in the manner shown in the following table.

Variable	Range	Quantisation
Position of the cart	–35 to +35 inches	5
Velocity of the cart	–30 to +30 inch/second	3
Angle of the pole	–12 to +12°	5
Angular velocity	–24 to +24°/second	3

Two of these variables are quantized into five values and two into three values. This means that, in four-dimensional space, there are 5 × 5 × 3 × 3 = 225 compartments. Figure 8.2 shows a two-dimensional section through the state space, showing the axes corresponding to the position of the cart and the angle of the pole. The state of the cart and pole at any time can be represented as a point in this four-dimensional state space. As the cart and pole moves, so the state changes, forming a path through this state space. As the state changes, it enters different boxes or compartments in the state space, and in each compartment a decision is made as to whether the cart gets pushed to the right or to the left. When the state goes beyond the ranges defined in the previous table, the system is said to have failed.

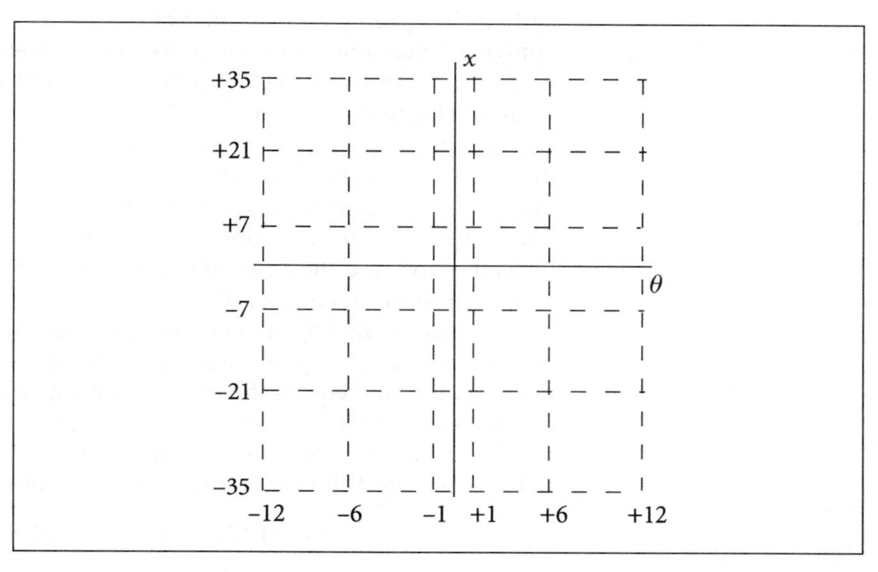

Figure 8.2 A 2D section through the state space

The following is a simplified description of the decision rules (Barto *et al.*, 1983). A global monitor or 'demon' inspects the incoming signals and selects the appropriate compartment. In each compartment there is a local decision maker also called a demon. It is the local demon's job to decide whether to ask for a right force or a left force. It does this by examining the two 'weights' associated with right and left decisions, w_r and w_l, and making a decision depending on which value is largest.

Initially, all weights are set to random values. When a decision is made, a count is kept of how long the pole remains balanced after that decision. When the pole eventually falls over, this is regarded as a complete system failure and the weights are modified. The values of the weights are the average of all actual times that the pole remained balanced after making a decision. So, after a failure, the count is used to adjust the weight in a box that was originally the larger of the two. For example, if the weights in a particular box are $w_r = 24$ and $w_l = 13$ after making three decisions to apply a force to the right ($c_r = 3$) and four decisions to apply a force to the left ($c_l = 4$), when the box is entered a decision will be taken to apply a force to the right. If the system then continues to balance the pole for a further count of 15 before failure, the value of w_r becomes:

$$w_r(c_r + 1) = \frac{w_r(c_r) \times c_r + \text{count}}{c_r + 1} = \frac{24 \times 3 + 15}{4} = 21.75$$

It is claimed that this method allowed the pole to be balanced for up to an hour after 60 hours of training.

8.3 Reinforcement learning

The relative success of Michie's boxes to control difficult non-linear systems has prompted a number of researchers to look at ways of applying better-known neural networks, such as the multi-layered perceptron (Anderson, 1989) to the same problem. Figure 8.3 shows one solution (Barto *et al.*, 1983).

The four variables from the actual cart and pole system are decoded into 162 binary variables, so that when the system is in the equivalent of one of Michie's boxes, only one of the variables x_i is 1. There are 162 'boxes' because the cart's position is quantized into only three regions instead of five, but the angle of the pole is quantized into six regions, the extra one being an angle of 0°.

The object, called the associative search element (ASE), is like a neuron which produces a weighted sum of the inputs. Since only one input is 1, the weighted sum equals the weight on the active input. However, random noise is added so that, when the system starts, the weights are zero but the sum has an equal probability of being positive or negative.

There is a hard-limiter on the output so that the final output, y, is:

$$y = +1 \text{ (Force to the right) if } w_i \geq 0$$

$$y = -1 \text{ (Force to the left) if } w_i < 0$$

The single weight acts like the difference between the two weights, w_r and w_l, in Michie's boxes. So, from an operating point of view, the two

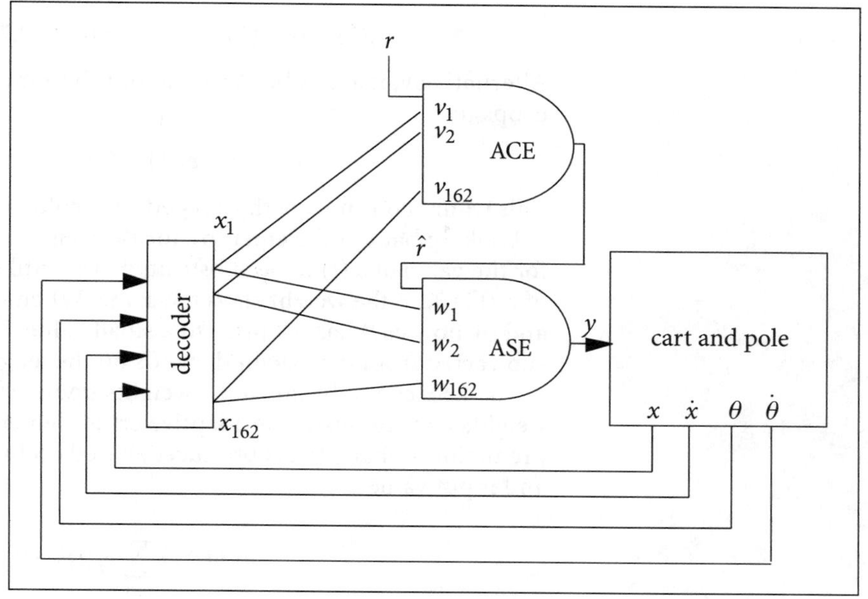

Figure 8.3 Neural network solution to the cart and pole problem

systems are the same. The difference comes in the training of the weights. To start with, assume that the system consists only of the ASE, and the reinforcement signal, $r(t)$ is applied directly to the ASE. The equation for the weight adjustment is:

$$w_i(t+1) = w_i(t) + \alpha r(t) e_i(t)$$

where α is a small positive constant, $r(t)$ is the reinforcement at time t, and $e_i(t)$ is the eligibility at time t. The reinforcement, $r(t)$, has a value of 0 during training until the system fails, at which point its value changes to –1. The eligibility is a function given by the equation:

$$e_i(t+1) = \delta e_i(t) + (1-\delta) y(t) x_i(t)$$

where δ lies between 0 and 1.

The eligibility performs a similar function to the count in Michie's boxes. When a box is entered, the eligibility for that box starts to rise exponentially, and when the system leaves the box, the eligibility drops exponentially. So the longer the system stays in a box, the larger the eligibility for that box, but the longer the time between leaving the box and eventually failing, the smaller the value of the eligibility for that box. Boxes with high eligibility therefore correspond to boxes that the system was in just before failure.

When a particular box is entered, x_i for that box is 1 while all other x_is are 0. An output decision is made, so y is either +1 or –1. Initially, if this is the first time that a box has been entered, $e_i(0) = 0$, so $e_i(1) = (1-\delta)y(0)$, which is either $+(1-\delta)$ or $-(1-\delta)$. For example, assuming that $y = +1$ and $\delta = 0.7$, then $e_i(1) = 0.3$. Next, the cart moves. If it is still in the same box, the eligibility becomes:

$$e_i(2) = \delta\, e_i(1) + (1 - \delta) = 0.3 \times 0.7 + 0.3 = 0.51$$

Alternatively, if a new box is entered, x_i becomes 0, and the eligibility drops to:

$$e_i(2) = \delta\, e_i(1) = 0.3 \times 0.7 = 0.21$$

This would continue, with $e_i(t)$ getting smaller and smaller.

Looking back at the equation for the weight adjustment, it is possible for the value of $e_i(t)$ to be substituted but, until the system fails, the value of $r(t)$ is 0, so the weight stays the same. When the system fails, $r(t) = -1$, and by now $e_i(t)$ has dropped to a small value. So the amount that is subtracted from the weight depends on the length of time before failure.

The ASE can only adjust the weights upon failure. In order to adjust the weights continuously, the adaptive critic element (ACE) is used. In this, a prediction signal, $p(t)$, is produced, based on the current set of weights and input values.

$$p(t) = \sum_{i=1}^{n} v_i(t)x_i(t)$$

This is a prediction of the final amount that will be deducted from the weight when the system fails. It produces an estimate of the reinforcement signal, $\hat{r}(t)$, as its output.

$$\hat{r}(t) = r(t) + \gamma p(t) - p(t-1)$$

where γ is between 0 and 1.

It is possible for $\hat{r}(t)$ to be positive when $r(t)$ is 0 and:

$$\gamma p(t) - p(t-1) > 0$$

This happens when the system moves from one box, which predicts a large negative value, to a box which predicts a smaller negative value. In other words, the system has entered a box which predicts a longer time before failure than the previous box.

The weights are continuously being adjusted using the equation:

$$v_i(t+1) = v_i(t) + \beta\hat{r}(t)q_i(t)$$

The variable $q_i(t)$ is called the trace and plays the same role as the eligibility in the ASE, the difference being that it is independent of the value of the output. It is found using the equation:

$$q_i(t+1) = \lambda q_i(t) + (1 - \lambda)x_i(t)$$

where λ is between 0 and 1.

The weights stop changing when $\hat{r}(t) = 0$, which happens when:

$$\gamma p(t) - p(t-1) = 0$$

Usually, γ is made to be close to 1 (0.95 say), so that the weights stop changing when the predictions are roughly the same.

It is claimed that this method is superior to Michie's boxes, as the pole can be made to balance for much longer and training is much faster.

This has been quite a detailed description of the way that the ACE and ASE operate. The main point, however, is that a method has been presented which allows a system with no model available to be controlled using only a failure signal. The ASE can be described as an action network that provides the control signal, and the ACE can be described as an evaluation network that allows continuous training (Anderson, 1989). This seems to be a very useful control strategy for these types of systems.

8.4 ADALINE

Neural networks can be trained to mimic the behaviour of a controller. This could be a linear controller when a linear model of the system is known, or a human controller when a model is not known (Widrow, 1987). A controller that is often used for the cart and pole is one which has proportional and derivative control. This means that the force used to control the trolley is proportional to the error and the derivative of the error between the actual output and the desired output. This controller is found using a simplified linear model of the cart and pole, and applying standard control theory for linear systems. Figure 8.4 shows the feedback control system.

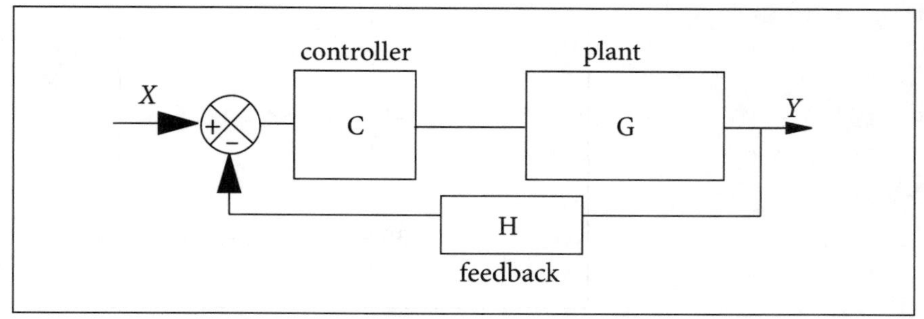

Figure 8.4 Feedback control system

Since the system is trying to control both the position of the cart and the angle of the pole, the 'output' of the system is a combination of both of these variables. In addition, a control strategy that is often used is referred to as 'bang-bang' control because the controller's output is its maximum value positively or negatively. A theory exists, called Pontryagin's maximum principle, that says that bang-bang control is time-optimal, which means that the system gets to where it wants to be in the shortest possible time (Widrow and Smith, 1964). So it is possible to design a controller using proportional and derivative action, and then simply put a hard-limiter on the output of the controller.

The controller can therefore be described by the equation:

$$\text{output} = +F_{\max} \text{ when } w_1\dot{\theta} + w_2\theta + w_3\dot{x} + w_4 x \geq 0$$

$$\text{output} = -F_{\max} \text{ when } w_1\dot{\theta} + w_2\theta + w_3\dot{x} + w_4 x < 0$$

This equation describes the action of a single neuron such as the ADALINE where the neuron fires (+1 output) when the weighted sum of its input is greater than some threshold (in this case 0), otherwise it does not fire (−1). So, a single neuron can be used to control the trolley and pole.

Comparing these equations with the description of the state space given in the section on Michie's boxes, it can be seen that this equation corresponds to a hyperplane through the state space. In a two-dimensional space, this corresponds to a straight line, as shown in Figure 8.5. Above the line, the output force is positive, and below the line the output force is negative.

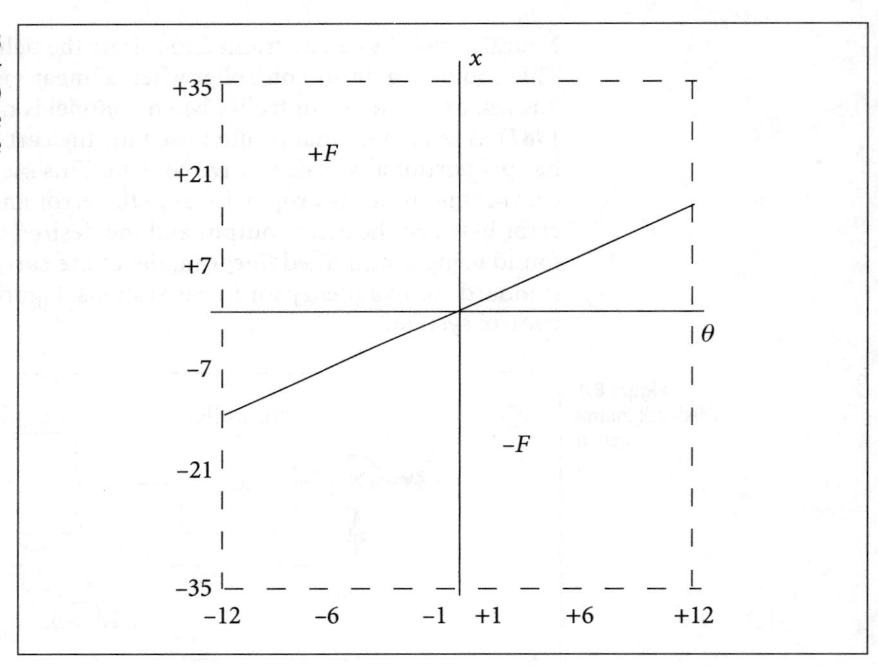

Figure 8.5 ADALINE controller represented in a 2D section through the state space

The weights are found by training using the delta rule (described in Chapter 2). The ADALINE monitors the system outputs (such as the angle of the pole and position of the cart) and the corresponding control action taken by the human. By pairing the system outputs with the controller output, a training set can be constructed where the present set of system outputs are the inputs to the ADALINE, and the controller output is the desired output of the ADALINE.

Clearly then, the best that this solution will ever achieve is to be able to learn to control a system as well as an existing controller (human or otherwise). The only situation where this is advantageous is where a system is currently controlled manually and has to be automated, but little is known about the system. The only way to automate it is to mimic the human controller, and the ADALINE does just this.

A major handicap of the single neuron is its inability to emulate all input-to-output relationships. This means that it might not be able to mimic a controller over the full range of circumstances. The multi-layer perceptron was shown in an earlier chapter to have an advantage over the ADALINE, which is that it can mimic any input/output relationship. Thus the multi-layer perceptron can be used to emulate any existing controller.

8.5 Multi-layered perceptron

In this section, three applications of the multi-layered perceptron will be examined. However, it must be noted that there are many more applications of this very versatile network (Miller *et al.*, 1990).

8.5.1 System identification

The property of a multi-layer perceptron network that is most interesting is its ability to emulate any input/output relationship. In a linear system this relationship is the transfer function, G, and the goal in system identification is to find an approximation of the system's transfer function, $\hat{G}$. This can be used for modelling or emulating the plant, or is sometimes used in adaptive control, where the effects of the controller outputs need to be known before they are applied to the plant (Miller *et al.*, 1990).

There are a number of ways that this can be done, but we will just look at one, shown in Figure 8.6.

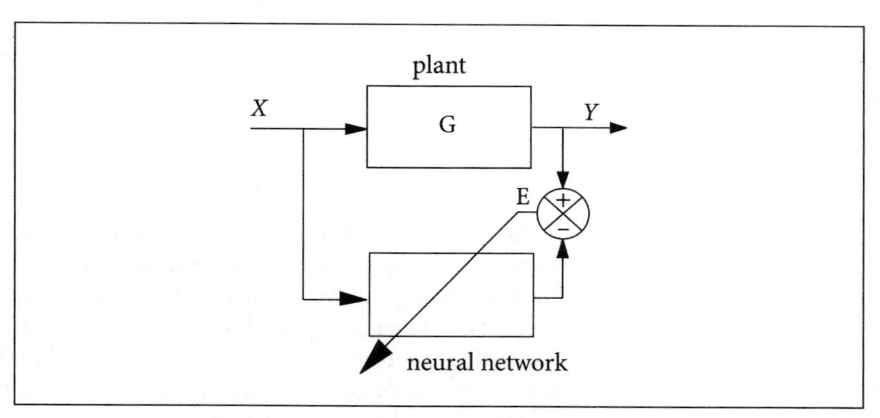

Figure 8.6 Training a network to emulate *G*

A training set of data is produced by sending inputs, x, to the system and recording the values of x and the corresponding values of the outputs, y. Note that the input and output could include the derivatives or they could include past values. A neural network is then trained by setting the inputs of the network to x and making the desired output of the network y. An error, e, is produced between the output of the network and the output of the plant, which can be used in back-propagation to adjust the weights. Once it has trained it effectively represents the transfer function of the system.

One problem with this method is deciding how to select the training data. The inputs and outputs should be representative of the states that the system will enter, but it can be difficult to know in advance how to do this.

8.5.2 Open-loop control

The ideal transfer function for a system is 1. If the transfer function of a plant is G, then from a control engineer's point of view, having the transfer function $1/G$ (called the inverse of G) would be very useful. If a controller had the form $1/G$, and this was placed in front of the plant, without feedback, the overall transfer function would be $1/G \times G = 1$. So, as a general principle, if the inverse of G can be approximated, $1/\hat{G}$, good control can be achieved using open-loop control. Since a multi-layered neural network can approximate any relationship, it should be able to find $1/\hat{G}$. Figure 8.7 shows the principle.

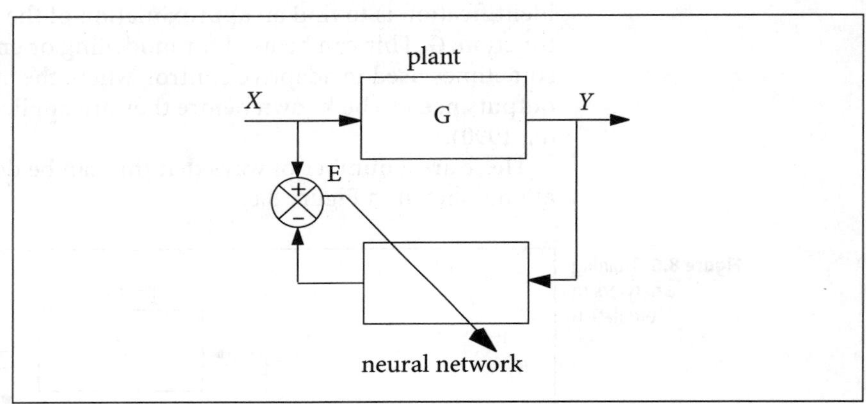

Figure 8.7 Training a network to emulate $1/G$

Essentially, the same method is used as for system identification. This time, a training set of data is produced by sending inputs, x, to the system and recording the values of x and the corresponding values of the outputs, y, (including the derivatives). A neural network is then trained by setting the inputs of the network to y and making the desired output x. Once it has trained, it effectively represents the inverse transfer function of the system being controlled.

8.5.3 Reinforcement learning using ADALINEs

A similar approach to the ASE and ACE method described earlier can be applied to the ADALINE if a large number of layers are used (Miller *et al.*, 1990). Although these neurons are described as ADALINEs, back-propagation is used to train the network, so it can be assumed that they are multi-layered perceptrons. First, a network is trained to emulate the plant, as described in this section. Then a neural network controller is attached to the emulator to make a single layer. This layer is then duplicated many times, as shown in Figure 8.8(a).

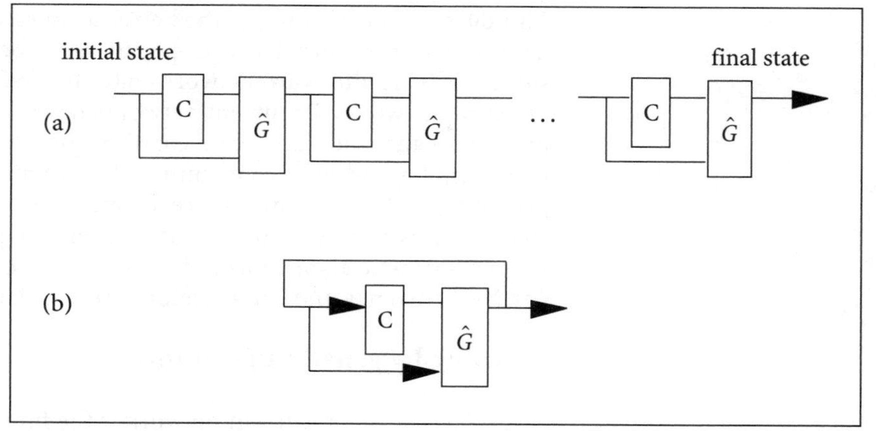

Figure 8.8 Layers of ADALINEs

The idea is that the initial state of the system is fed in to the first layer. The controller drives the emulator to produce an output that is the next state of the system. This is fed to the next layer, which repeats this process, passing on its output to the next layer. This goes on until the system reaches a state that corresponds to failure. The final state is compared with the desired goal state and an error produced, which is used to back-propagate weight changes in each of the controllers.

All of the controllers in each layer are identical, so instead of having a large number of identical layers, the output could be fed back to the input of a single layer, as shown in Figure 8.8(b). The error is therefore propagated back through the network and cycles around, the number of times being equal to the number of states between the initial and final states.

Although this method can produce successful control of quite complex systems, the number of cycles of weight changes can be very large. However, the notion of recurrence has been introduced. This will be explored in the next section.

8.6 Recurrent neural networks

A recurrent neural network is one where feedback is allowed, so recurrent networks are a particular type of feedback network. In Chapter 5 the Hopfield network was described, which was also a feedback network. The differences between the recurrent networks in this section and the Hopfield networks are:

- In a Hopfield network, inputs are applied once at the start, and then disconnected. In recurrent networks inputs arrive at each iteration.
- In the Hopfield network, the neurons update asynchronously. In recurrent networks the neurons update synchronously, which means that they all fire together.

- Training in the Hopfield network involved a single calculation. In recurrent networks training is more complex, and involves many iterations.

This latter point, training, is the key issue in research into recurrent neural networks. Since back-propagation has been demonstrated so successfully, feedforward networks have tended to dominate in the field of neural networks. Recurrent networks have not been so successful, mainly because the training mechanisms that have been proposed are very complex and time-consuming. The two major approaches are back-propagation through time and real-time recurrent learning (Miller *et al.*, 1990). In this section, two alternative approaches will be examined. The first uses genetic algorithms to find a set of weights. The second uses partial recurrency and can therefore use standard back-propagation.

8.6.1 Learning by genetic algorithms

A special case of a recurrent network is the fully connected one, where the output of each neuron is connected to the input of every other neuron and itself. Figure 8.9 shows a fully connected recurrent network with three neurons and two inputs.

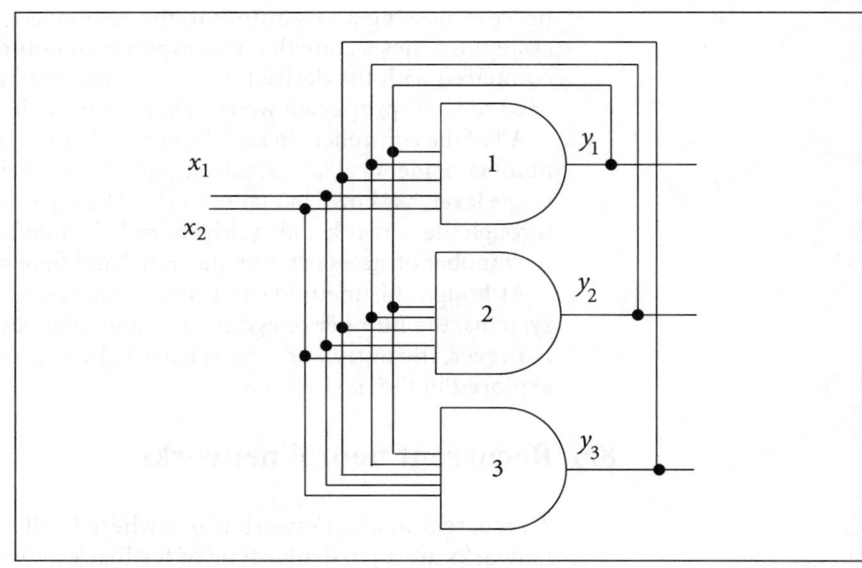

Figure 8.9 Fully connected recurrent neural network

This type of network has been used to control the cart and pole (Wieland, 1991). The network consisted of six neurons. The weights are found by training a population of controllers and selecting the best ones. A genetic algorithm is used to 'breed' new controllers from the best of the current population. There is not enough room in this book to go into details of genetic algorithms, and the interested reader would be advised to find a book that deals with them exclusively (Davis, 1991). However, briefly, the weights are coded as binary strings in an analogy with chromosomes. Breeding consists of swapping parts of a pair of chromosomes to produce new chromosomes. Mutation is also allowed to introduce random variations. In this way, the genetic algorithm ensures that 'good' chromosomes are preserved and that new variations are generated.

A set of weights can be quickly found using this method. Thus this method provides a general solution to finding controllers for arbitrary systems. The architecture of the network allows the dynamic as well as static nature of the plant under control to be captured, and the genetic algorithm allows the weights to be found.

8.6.2 Elman nets

Elman nets (Elman, 1990) are becoming increasingly popular in system modelling and prediction. Earlier, it was said that multi-layered perceptrons are useful for system identification because they can capture the relationship between the inputs and outputs of a system. In a linear system this would be described as the transfer function of the system. However, training consists of showing the network a set of input and output pairs of data, with no consideration given as to how the system arrived at those data. Thus the data, and the resultant model, represent only the static model of the system. Of more use to a control engineer is the dynamic model of the system, which takes into account the way in which the system changes from one state to the next. Feedforward networks cannot capture this information, but recurrent networks can.

The Elman network has been proposed as such a recurrent network for system dynamic modelling (Pham and Liu, 1993). A typical network is shown in Figure 8.10, which shows a network with two inputs, two outputs and three neurons in the hidden layer. The number of neurons in the hidden layer is duplicated in the context layer.

The Elman network is a three-layer network, but only its middle layer is recurrent. Its advantage over fully recurrent networks is that back-propagation can be used to train the network. This is because the connections to the context units are fixed, so that the context units behave as a delay of one sampling period.

One interpretation of this network is that the outputs of the hidden layer represent the state of the network. The outputs of the network are functions of the present state, the previous state (as supplied by the context units) and the present inputs. This means that when the network is shown a set of inputs, it can learn to give the appropriate outputs in the context of the previous states of the network.

The Elman network was originally developed for speech recognition, but there is growing interest in other areas, such as system identification and short-term prediction, which is used in motion planning for robotics (Meng and Picton, 1993).

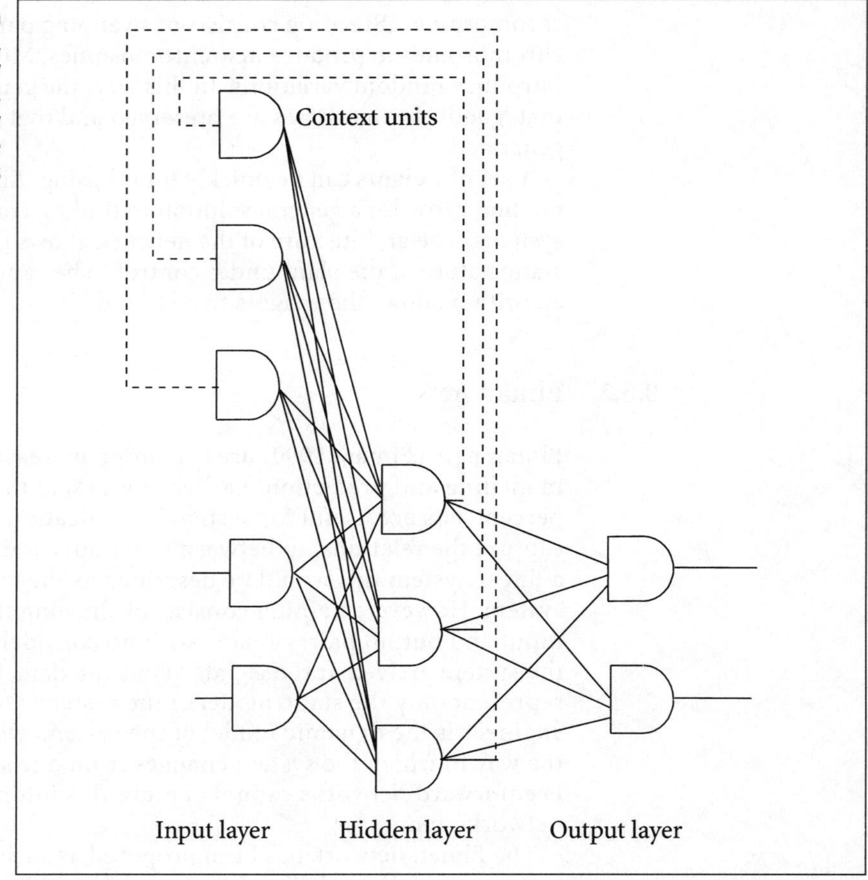

Figure 8.10 An Elman network. Dashed lines show fixed connections, and solid lines show adjustable connections

8.7 The Kohonen network

Kohonen networks were described in Chapter 7. They are self-organising networks which can be used as pattern classifiers. When acting as a pattern classifier, input patterns that are similar produce output responses in neurons that are physically close together, and input patterns that are dissimilar produce output responses in neurons that are physically far apart.

Kohonen networks can also be used in control, in problems such as the cart and pole (Ritter *et al*, 1992). One layer of neurons is used, but with two sets of weights: an input set and an output set. The input set is trained on the input data, and the output set is trained on the corresponding output data, as shown in Figure 8.11.

Recall that training consists of adapting the weight of the neuron that is most similar to the input pattern using Kohonen learning. So, in this example, the input weights on the winning neuron are modified using Kohonen learning so that they eventually self-organise. This means that states that are close to each other in the state space cause neurons that are physically close together to fire. The output weights in the same winning

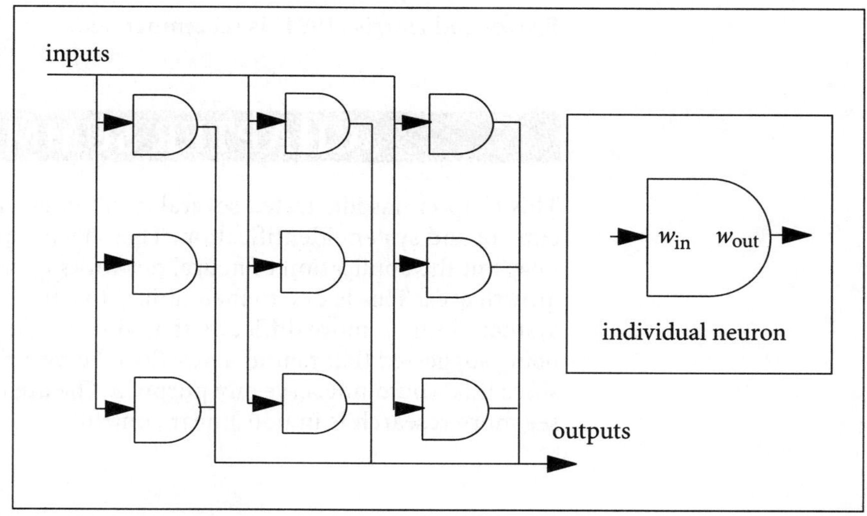

Figure 8.11
Kohonen networks

neuron adjust so that they become more like the appropriate output response, but they do not self-organise, as they are inextricably linked to the input weights.

The system can be trained using supervised or unsupervised learning. During supervised learning, the network is shown input and output pairs taken from a known controller. The output force is a mixture of the known output force and the learnt output force. Initially, the known output force is used, but as training progresses, the learnt force takes over. It is reported that after training a 25 × 25 array of neurons for 1000 iterations the network successfully emulates the controller.

In unsupervised learning, since the 'correct' force to apply is not known, a reward function is used to indicate when the system is in a 'good' state. The reward function used is $R(\theta) = -\theta^2$. Training is inevitably more complex and therefore takes longer, but the system can be controlled after only 3000 iterations.

8.8 Neuro-fuzzy systems

One of the major growth areas in research in control has been in fuzzy logic control (Picton, 1997). This has proven to be practical and effective, but has one major drawback. Essentially, a fuzzy logic controller uses a set of rules that determine how it behaves, and the inputs and outputs use fuzzy values which are derived by converting actual inputs and outputs through fuzzy membership functions – a process of fuzzification for inputs and defuzzification for outputs. The choice of rules and the shape and position of the membership functions is predetermined, and is usually a best estimate from the data. Ideally, a system should be able to learn these from training data. Research is therefore ongoing in the area of training neural networks to learn a particular relationship between the inputs and the outputs, and to extract from a trained network the set of rules and the membership functions. For further reading the book by

Brown and Harris (1994) is recommended.

CHAPTER SUMMARY

This chapter has illustrated several techniques that have been used in control and system identification. These techniques are still relatively new, but the application of neural networks in control seems to be a growth area. This is due to their ability to control complex non-linear systems with no more difficulty than linear systems. However, it is not being suggested that neural networks take over from linear controllers, since this would not serve any purpose. The area where we will probably see more research is in non-linear control

SELF-TEST QUESTIONS

1 In Michie's boxes applied to the cart and pole problem a particular box has a weight $w_r = 21.75$ and a weight $w_l = 13$. The box has been visited eight times, and has made a total of four decisions to apply a force to the right. This time it also decides to apply a force to the right, which results in the pole staying upright for a count of 10. What is the new value of the weight w_r?

2 The ASE in a reinforcement learning system uses eligibility as a measure of the influence that a decision has on the performance of a system. Assuming that the eligibility starts from zero, and that the value of δ is 0.7 and that the output is +1:

(a) what are the values of the eligibility if the system stays in a particular box for a count of 4?

(b) what are the values of the eligibility for a count of 4 after leaving the box?

3 An ADALINE is set to control a system. The inputs to the controller are a variable x and its derivative. The output if $+Y$ if the weighted sum is greater than 0 and $-Y$ if it is less than or equal to zero. The decision surface or hyperplane passing the through the state space is given by the following equation:

$$\dot{x} = 1.5x + 5$$

What are the values of the weights in the ADALINE?

4 A system is defined by the following difference equation:

$$Y_k = 0.2Y_{k-1} + 0.3X_k$$

(a) Show how this could be simulated using a recurrent neural network.

(b) Set k to 3 and expand the equation so that it contains the values of X from X_3 to X_0. If we ignore any weights less than 0.01, show how the same equation can be simulated using previous input values in a feedforward network.

SELF-TEST ANSWERS

1 The equation used to update the weight is:

$$w_r(c_r + 1) = \frac{w_r(c_r) \times c_r + \text{count}}{c_r + 1}$$

Substituting for the values given in the question:

$$w_r(c_r + 1) = \frac{21.75 \times 4 + 10}{4 + 1} = \frac{97}{5} = 19.4$$

2 (a) With $\delta = 0.7$, the equation for eligibility is given by the following when the system is within the box so that $x_i = 1$:

$$e_i(t + 1) = 0.7e_i(t) + 0.3$$

Starting from 0, the eligibility goes:

$$e_i(0) = 0$$

$$e_i(1) = 0.7 \times 0 + 0.3 = 0.3$$

$$e_i(2) = 0.7 \times 0.3 + 0.3 = 0.51$$

$$e_i(3) = 0.7 \times 0.51 + 0.3 = 0.66$$

$$e_i(4) = 0.7 \times 0.66 + 0.3 = 0.76$$

(b) The system now leaves the box, so x_i becomes 0 and the eligibility is:

$$e_i(t + 1) = 0.7e_i(t)$$

So for the next four counts the values are:

$$e_i(4) = 0.7 \times 0.76 = 0.53$$

$$e_i(5) = 0.7 \times 0.53 = 0.37$$

$$e_i(6) = 0.7 \times 0.37 = 0.26$$

$$e_i(7) = 0.7 \times 0.26 = 0.18$$

3 The equation for the hyperplane can be rearranged to give:

$$\dot{x} - 1.5x - 5 = 0$$

This corresponds to weights of 1 for the derivative of x, –1.5 for x and a weight of –5 for the bias which has a constant +1 input.

4 (a) The system could be set up so that the output, Y, is fed back as a second input with a weight of 0.2, whereas the current input has a weight of 0.3. The inputs are read in to the neuron synchronously, so the fed-back value can be regarded as the previous output.

(b) The equation can be expanded as follows:

$$
\begin{aligned}
Y_3 &= 0.2Y_2 + 0.3X_3 = 0.2(0.2Y_1 + 0.3X_2) + 0.3X_3 \\
&= 0.04Y_1 + 0.06X_2 + 0.3X_3 \\
&= 0.04(0.2Y_0 + 0.3X_1) + 0.06X_2 + 0.3X_3 \\
&= 0.008Y_0 + 0.012X_1 + 0.06X_2 + 0.3X_3 \\
&= 0.008(0.2Y_{-1} + 0.3X_0) + 0.012X_1 + 0.06X_2 + 0.3X_3 \\
&= 0.0016Y_{-1} 0.0024X_0 + 0.0012X_1 + 0.06X_2 + 0.3X_3
\end{aligned}
$$

Ignoring any terms less than 0.01 gives:

$$Y_3 = 0.012X_1 + 0.06X_2 + 0.3X_3$$

So we can generate the new output value using the current input value plus the previous two, all suitably weighted. The solution is therefore a single neuron with three inputs weighted with 0.3 for the current input, 0.06 for the previous input and 0.012 for the input value before that.

CHAPTER 9

Threshold logic

CHAPTER OVERVIEW

This chapter gives describes the specific application of neural networks to the problem of representing Boolean logic functions. It explains:

- how a single neuron can represent linearly separable logic functions
- how to test for linear separability
- how all logic functions, even non-linearly separable ones, can be classified
- how non-linearly separable logic functions can be implemented using a single neuron with multiple outputs.

9.1 Introduction

A great deal of research has gone into threshold logic, with some useful theoretical results (Dertouzos, 1965; Muroga, 1971; Hurst, 1978). The main aim of threshold logic is to take what are essentially McCulloch–Pitts neurons, and use them to represent completely specified binary logic functions.

Some differences are:

- inputs and outputs are 0,1 binary
- weights can only have integer values – positive or negative
- the output function is a hard-limiter
- the offset, or threshold, is also a positive or negative integer

Figure 9.1 shows the usual way of depicting a threshold logic gate. The output is defined as:

$$y = 1 \quad \text{if} \geq t$$

$$y = 0 \quad \text{otherwise}$$

So, a threshold logic gate, or network of gates, is used to implement a specified logic function. There is no need for generalisation, as the inputs and corresponding outputs are all known.

Figure 9.1
Threshold logic gate

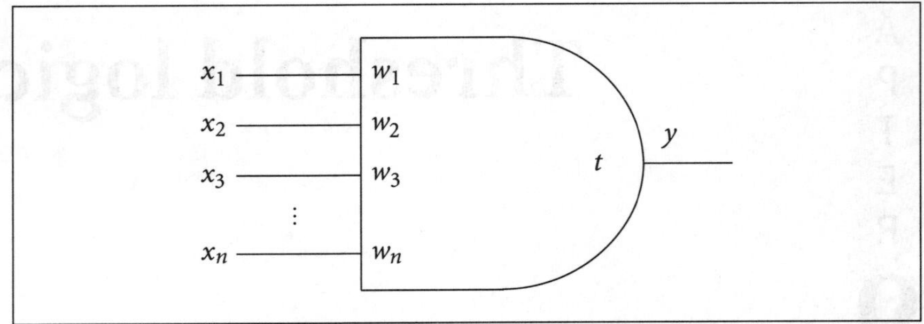

One way of establishing the weights is to use supervised learning, but again it must be emphasised that finding weights that produce an incorrect output even once is regarded as a failure.

Attempts have been made to find a theoretical way of selecting the weights. One area of research is finding ways to determine whether or not a function is linearly separable. Another area is the classification of logic functions.

9.2 A test for linear separability

Over a period of time there have been three tests developed in this area. The first test is called the test for unateness. A function is unate if, in the reduced Boolean expression for the function, none of the input variables appears in both uncomplemented and complemented form. So, for example, the function:

$$f(x) = (x_1 \wedge x_2) \vee (x_1 \wedge x_3)$$

is unate, whereas the function:

$$f(x) = (x_1 \wedge x_2) \vee (\neg\, x_1 \wedge x_3)$$

is not unate because x_1 appears in the expression in both complemented ($\neg\, x_1$) and uncomplemented (x_1) form.

It has been shown (Hurst, 1978) that all linear separable functions are unate. Unfortunately, not all unate functions are linearly separable. For example, the function:

$$f(x) = (x_1 \wedge x_2) \vee (x_3 \wedge x_4)$$

is not linearly separable, even though it is unate.

The second test is monotonicity. In order to explain this, comparability needs to be defined. Two functions, $f_1(x)$ and $f_2(x)$, are comparable if all of the 1s in $f_1(x)$ have corresponding 1s in $f_2(x)$ for the same values of x, or if all of the 1s in $f_2(x)$ have corresponding 1s in $f_1(x)$ for the same values of x. This can be expressed as:

for all combinations of input, $f_1(x) \geq f_2(x)$ or $f_1(x) \leq f_2(x)$

The following tables show two examples where $f_1(x)$ and $f_2(x)$ are comparable functions. In the first case, the functions are comparable because $f_1(x) \leq f_2(x)$ for all values of x. In the second case the two functions are comparable because $f_1(x) \geq f_2(x)$ for all values of x.

x_1	x_2	x_3	$f_1(x)$	$f_2(x)$
0	0	0	0	0
0	0	1	0	0
0	1	0	1	1
0	1	1	1	1
1	0	0	0	1
1	0	1	1	1
1	1	0	0	0
1	1	1	1	1

x_1	x_2	x_3	$f_1(x)$	$f_2(x)$
0	0	0	0	0
0	0	1	0	0
0	1	0	0	0
0	1	1	1	0
1	0	0	1	1
1	0	1	0	0
1	1	0	1	1
1	1	1	1	0

If a function is decomposed about one of its input variables, x_i, two disjoint functions result:

$$f(x_1, x_2, ..., x_i = 0, ..., x_n) \text{ and } f(x_1, x_2, ..., x_i = 1, ..., x_n)$$

These two functions are monotonic if they are comparable. Furthermore, the function itself is said to be 1-monotonic if it decomposes into comparable functions for each input variable x_i, for $i = 1$ to n.

Taking this further, a function is said to be 2-monotonic if it can be decomposed about pairs of input variables and still produce comparable functions. Still further, a function is said to be k-monotonic if it can be decomposed about k variables and still produce comparable functions.

It can be shown that all linearly separable functions are completely monotonic, that is, they can be decomposed about all combinations of their input variables and still produce comparable functions. Take an example, where a threshold function has weights of 1, 2, and −1 and a threshold of 2. The function is shown in the following table and, since it has been implemented using a single threshold logic gate, it must be linearly separable.

x_1	x_2	x_3	*net*	$f(x)$
0	0	0	0	0
0	0	1	−1	0
0	1	0	2	1
0	1	1	1	0
1	0	0	1	0
1	0	1	0	0
1	1	0	3	1
1	1	1	2	1

The function can be written as a Boolean expression:

$$f(x) = (x_2 \wedge \neg x_3) \vee (x_1 \wedge x_2)$$

The first thing to note is that the function is unate, so it passes the first test. Next test for 1-monotonicity.

1-monotonicity

	$x_i = 0$ (f_0)	$x_i = 1$ (f_1)	comparability
x_1	$f(0, x_2, x_3) = x_2 \wedge \neg x_3$	$f(1, x_2, x_3) = x_2$	$f_1 \geq f_0$
x_2	$f(x_1, 0, x_3) = 0$	$f(x_1, 1, x_3) = \neg x_3 \vee x_1$	$f_1 \geq f_0$
x_3	$f(x_1, x_2, 0) = x_2$	$f(x_1, x_2, 1) = x_1 \wedge x_2$	$f_0 \geq f_1$

For each input variable, decomposition produces two comparable functions. So the function is 1-monotonic.

2-monotonicity

	f_{00} f_{10}	f_{01} f_{11}	comparability
x_1x_2	$f(0, 0, x_3) = 0$	$f(0, 1, x_3) = \neg x_3$	
	$f(1, 0, x_3) = 0$	$f(1, 1, x_3) = 1$	$f_{11} \geq f_{01} \geq f_{10} \geq f_{00}$
x_1x_3	$f(0, x_2, 0) = x_2$	$f(0, x_2, 1) = 0$	
	$f(1, x_2, 0) = x_2$	$f(1, x_2, 1) = x_2$	$f_{11} \geq f_{10} \geq f_{00} \geq f_{01}$
x_2x_3	$f(x_1, 0, 0) = 0$	$f(x_1, 0, 1) = 0$	
	$f(x_1, 1, 0) = 1$	$f(x_1, 1, 1) = x_1$	$f_{10} \geq f_{11} \geq f_{01} \geq f_{00}$

For each pair of input variables, decomposition produces four comparable functions. So the function is 2-monotonic.

3-monotonicity

$x_1x_2x_3$	$f(0, 0, 0) = 0$	$f(0, 0, 1) = 0$
	$f(0, 1, 0) = 1$	$f(0, 1, 1) = 0$
	$f(1, 0, 0) = 0$	$f(1, 0, 1) = 0$
	$f(1, 1, 0) = 1$	$f(1, 1, 1) = 1$

comparability $\quad f_{111} \geq f_{110} \geq f_{010} \geq f_{101} \geq f_{100} \geq f_{011} \geq f_{001} \geq f_{000}$

Every function is 3-monotonic, so this function is completely monotonic.

It was thought for some time that this was a universal test for linear separability. However, a counter-example was found for $n = 9$ (Gabelman, 1962). It has been shown, however, that for n up to 8, complete monotonicity is proof of linear separability (Muroga *et al.*, 1970).

Finally, for $n \geq 8$, the test for k-asummability proves linear separability. Before asummability can be defined, summability needs to be defined.

An n-variable function, $f(x)$, is said to be k-summable if k true minterms and k false minterms can be found such that the vector summation of the true minterms equals the vector summation of the false minterms.

$$\sum^{\text{TRUE}} x_{ip} = \sum^{\text{FALSE}} x_{ip} = \text{ for } p = 1 \text{ to } k \text{ and } i = 1 \text{ to } n$$

The value of k is in the range 2 to 2^{n-1}.

Taking the previous example, select any two 1s and two 0s. For example, if the last four rows are taken, the sum of the minterms is:

False				True		
x_1	x_2	x_3		x_1	x_2	x_3
1	0	0		1	1	0
1	0	1		1	1	1
2	0	1		2	2	1

These two vectors are not equal, so the function appears not to be 2-summable. However, another set of minterms might be found that are 2-summable, so every combination has to be tried. If no combination can be found then the function is 2-asummable.

All linear separable functions are k-asummable, which means that for all values of k, no combination of minterms can be found such that their vector sums are equal (Elgot, 1960).

9.3 Classification of logic functions

Another way to determine linear separability and to select values for the weights is to classify logic functions. When presented with a new logic function, find the class that it belongs to. This will have a known solution, which can then be manipulated to give the desired solution.

The first method is to use the Chow parameters (Winder, 1971). These are calculated by correlating the function with all of the input variables. This can be done as a matrix operation by multiplying the input variable matrix with the function. The binary values of –1 and +1 are used in the calculation.

The example of the function $f(x) = (x_2 \wedge \neg x_3) \vee (x_1 \wedge x_2)$, given in the previous section, has the following Chow's parameters:

$$\begin{bmatrix} +1 & +1 & +1 & +1 & +1 & +1 & +1 & +1 \\ -1 & -1 & -1 & -1 & +1 & +1 & +1 & +1 \\ -1 & -1 & +1 & +1 & -1 & -1 & +1 & +1 \\ -1 & +1 & -1 & +1 & -1 & +1 & -1 & +1 \end{bmatrix} \begin{bmatrix} -1 \\ -1 \\ +1 \\ -1 \\ -1 \\ -1 \\ +1 \\ +1 \end{bmatrix} = \begin{bmatrix} -2 \\ 2 \\ 6 \\ -2 \end{bmatrix}$$

Notice that a row of all +1s is included. This corresponds to a constant +1 input variable, usually called x_0. Also note that this process corresponds to Hebbian learning as described in Chapter 2.

The Chow parameters, b_i, are useful in that they have similar properties to the weights, w_i, in the final threshold logic gate implementation. These properties are:

- If a Chow parameter, b_i, is negative then the corresponding weight, w_i, will be negative. Similarly, if b_i, is positive, the corresponding w_i, is positive.
- If a Chow parameter, b_i, is zero then the corresponding weight, w_i, will be zero.
- For any two Chow parameters, b_i and b_j, if one is greater than the other, in the final implementation the corresponding weight will be greater than the other. That is, if $b_i > b_j$, then $w_i > w_j$.
- If two parameters are equal, their corresponding weights will be equal. That is, if $b_i = b_j$, then $w_i = w_j$.

Tables have been produced of the Chow parameters of all linearly separable functions of up to seven variables. The following table shows all of the entries for the case where $n \leq 3$.

b_i				w_i			
8	0	0	0	1	0	0	0
6	2	2	2	2	1	1	1
4	4	4	0	1	1	1	0

Remarkable as it may seem, there are only three entries. This means that all linearly separable functions of three or fewer variables can be classified by these three sets of parameters. They are arranged in descending order of magnitude, which in this case is called the canonical form.

It is clear from the earlier example that the logic function belongs to the second class. The way that the values of the weights of a function are determined is to match the parameters with the canonical set for its class, and select the corresponding weights, adding a negative sign if the parameter is negative. So, for this example, the parameters are:

b_0	b_1	b_2	b_3
−2	2	6	−2

The weights are therefore:

w_0	w_1	w_2	w_3
−1	1	2	−1

The threshold, t, is calculated using the formula:

$$t = 0.5\left(\sum_{i=1}^{n} w_i - w_0 + 1\right) = 0.5(2 + 1 + 1) = 2$$

The Chow parameters are therefore very powerful because they not only determine whether a function is linearly separable or not, by there being a corresponding entry in the table, but also give us the optimum set of weights to implement the function. Their limitation, of course, is that they do not help if the function is not linearly separable.

An extension which shall briefly be mentioned is the use of the Rademacher–Walsh spectrum (Hurst, 1978). Essentially this is the same as the Chow parameters for the low order spectral coefficients, but also includes a number of higher order coefficients. These are found by correlating the function with the various exclusive-or combinations of input variables. For the three-variable example these are:

$$x_1 \oplus x_2$$

$$x_1 \oplus x_3$$

$$x_2 \oplus x_3$$

$$x_1 \oplus x_2 \oplus x_3$$

where $\oplus$ is the symbol for the exclusive-or function. So there are four extra coefficients, making eight in all. For n variables, the total number of coefficients is 2^n, and it has been shown that the 2^n basis functions represented by these coefficients form an orthogonal set (Hurst, 1978). The original function can be reconstructed from the coefficients using these basis functions. For example, a two-input function would be reconstructed as:

$$f(x_1,x_2) = \frac{1}{4}(R_0 x_0 + R_1 x_1 + R_2 x_2 + R_{1,2} x_1 \oplus x_2)$$

The Rademacher–Walsh spectrum is useful because it can provide a new classification scheme that includes non-linearly separable functions. Methods have been proposed which use the spectrum of a function to implement even non-linearly separable function using threshold logic gates, in a method called spectral translation (Hurst, 1978).

9.3.1 Higher-order neural networks

As long ago as 1963, the idea that neurons use the weighted sum of not just the inputs, but combinations of the inputs, was known (Klír and Valach, 1965). It has been suggested that if all of the combinations of inputs are included, the function can be completely specified by a weighted sum. For example, a two-input binary function can be written as:

$$f(x_1,x_2) = 1 \text{ if } w_0x_0 + w_1x_1 + w_2x_2 + w_{1,2}x_1x_2 > 0$$

$$f(x_1,x_2) = 0 \text{ if } w_0x_0 + w_1x_1 + w_2x_2 + w_{1,2}x_1x_2 \leq 0$$

The extra term, $w_{1,2}x_1x_2$, is a product of the two input variables. If it is assumed that the inputs have the values of +1 or −1, then the product is:

x_1	x_2	x_1x_2
−1	−1	+1
−1	+1	−1
+1	−1	−1
+1	+1	+1

An alternative interpretation of the product of the two input variables is the exclusive-or, which can be seen if 0 and 1 are used instead of −1 and +1. The weighted sum therefore corresponds to the Rademacher–Walsh spectrum.

Returning to neural networks, various authors have suggested using higher-order neural networks, which are networks that also include the product of the input variables, such as the 'sigma-pi' units (Rumelhart and McClelland, 1986). It is claimed that these higher-order networks perform better than ordinary neurons (Giles and Maxwell, 1987).

Higher-order networks, such as those made from sigma-pi units, can be trained in the same way as ordinary neurons using Hebbian learning. If an n-input logic function, which is completely known, is trained using Hebbian learning, the weights will be equivalent to the Rademacher–Walsh coefficients. This assumes that all possible combinations of input are used in the training, so that there would be 2^n weights. So, a single neuron with 2^n weights is all that is needed to represent the logic function.

If pattern recognition is now considered, where training uses a small sample of input and output data, could a single neuron with 2^n weights be trained, and would it generalise? The answer would appear to be that it would train, and that all of the patterns in the training set would give the correct response. However, the network would not generalise. Instead, it would give an output of 0 for all of the unseen patterns (Picton, 1991a). Reducing the number of weights gives back the ability to generalise, but no way is known of reducing the number of weights and still ensuring that the neuron will train.

It therefore seems that somewhere between multiple layers of ordinary neurons and a single higher-order neuron with 2^n weights there is an optimum network. As yet, no method exists for finding that optimal network.

9.4 Multi-threshold logic functions

To finish off this chapter, multi-threshold elements will be described (Haring, 1966). These are one way of implementing non-linearly separable functions. Essentially, a multi-threshold logic gate works in exactly the same way as a single-output threshold logic gate, except that there are several outputs. Each output has its own threshold, so that the weighted sum is compared with different values for each output. Figure 9.2 shows a three-output logic gate.

Figure 9.2 A three-output threshold logic gate

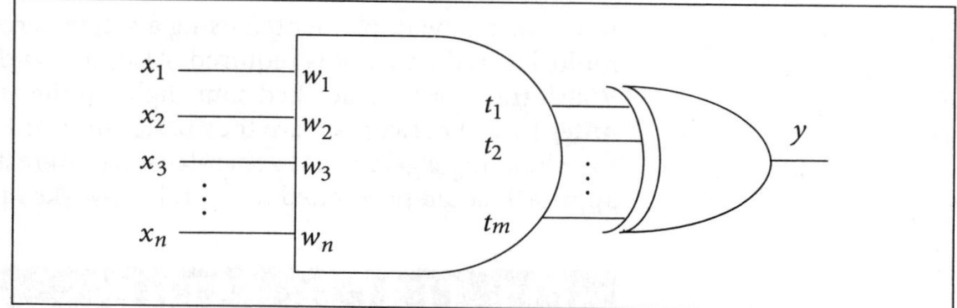

Figure 9.2 shows that the outputs are passed through an exclusive-or gate. If $t_1 < t_2 < t_3$, the final output, y, is:

$$y = 0 \text{ if } \sum_{i=1}^{n} w_i x_i < t_1 \text{ or if } t_2 \leq \sum_{i=1}^{n} w_i x_i < t_3$$

$$y = 1 \text{ if } t_1 \leq \sum_{i=1}^{n} w_i x_i < t_2 \text{ or if } t_3 \leq \sum_{i=1}^{n} w_i x_i$$

Although it can be shown that multi-threshold logic gates can implement any logic function (Hurst, 1978), the problem of finding suitable weights has never really been solved. One solution is to set weights which are powers of 2. This makes the value of *net* unique for each input pattern, and thresholds are set whenever the output changes from 0 to 1 or 1 to 0. Although this is a simple solution which needs no calculations, it usually produces a solution with far more thresholds than can be found by other means.

Just as the Chow parameters are used in a look-up table to find the weights for a single threshold function, so tables have been constructed for multi-threshold logic functions of four variables (Mow and Fu, 1968), which use the Rademacher–Walsh coefficients to classify the function.

So, although the multi-threshold logic gate is very versatile, it suffers from the problem of finding weights. One suggestion is to use the spectral translation method and then manipulate the network into a multi-threshold form (Picton, 1981). Very little research has gone into multi-threshold neural networks in general, but it would seem that this could be a potential area for development.

CHAPTER SUMMARY

This chapter has described some of the theoretical work that has been done on threshold logic. The threshold logic gates are McCulloch–Pitts neurons which have been applied to the problem of implementing logic functions as an alternative to Boolean logic gates such as AND gates. Although a great deal of effort went into threshold logic, it has never caught on as a viable means of building computers. However, many of the theoretical results can be applied to neural networks in general. For example, the test for linear separability could be used to determine if a function can be implemented using a single perceptron, or whether a multi-layered network is required. Also, the work on the Rademacher–Walsh transform could shed some light on the argument about higher-order neural networks – are they better or not? Finally, the multi-threshold logic gate could generate some interesting research if a similar approach could be applied to neural networks in general.

SELF-TEST QUESTIONS

1 The function $y = x_2 \lor x_1 \land \neg x_3$ is shown in the table below:

x_1	x_2	x_3	y
0	0	0	0
0	0	1	0
0	1	0	1
0	1	1	1
1	0	0	1
1	0	1	0
1	1	0	1
1	1	1	1

Use the test for k-asummability to show that the function is linearly separable.

2 Find the Chow's parameters for the function in Question 1 and hence the appropriate weights.

3 Show that the following function is not linearly separable:

x_1	x_2	x_3	y
0	0	0	0
0	0	1	0
0	1	0	0
0	1	1	0
1	0	0	1
1	0	1	0
1	1	0	0
1	1	1	1

4 Find the appropriate thresholds for a multi-threshold solution to the function given in question 3 when weights of 4, 2 and 1 are used for w_1, w_2 and w_3 respectively.

SELF-TEST ANSWERS

1 The 0s are at the minterms 000, 001 and 101. The 1s are at the minterms 010, 011, 100, 110 and 111. Using the 0 minterms, they can be treated separately for 1-asummability, in pairs for 2-asummability or all together for 3-asummability.

1-asummability	000	001	101
2-asummability	000	000	001
	<u>001</u>	<u>101</u>	<u>101</u>
	001	101	102
3-asummability	000		
	001		
	<u>101</u>		
	102		

Now try to find the same result using any combination of 1 minterms. It is impossible, so the function is linearly separable.

2 The Chow's parameters are found by multiplying the function with each of the inputs in turn, including a constant 1. But first all of the 0s must be converted to −1s.

$$\begin{bmatrix} +1 & +1 & +1 & +1 & +1 & +1 & +1 & +1 \\ -1 & -1 & -1 & -1 & +1 & +1 & +1 & +1 \\ -1 & -1 & +1 & +1 & -1 & -1 & +1 & +1 \\ -1 & +1 & -1 & +1 & -1 & +1 & -1 & +1 \end{bmatrix} \begin{bmatrix} -1 \\ -1 \\ +1 \\ +1 \\ +1 \\ -1 \\ +1 \\ +1 \end{bmatrix} = \begin{bmatrix} +2 \\ +2 \\ +6 \\ -2 \end{bmatrix}$$

Comparing with the table given in Section 9.3, the appropriate weights are 1, 1, 2, −1. The threshold is 0.5(1 + 2 − 1 − 1 + 1) = 1.

3 The most thorough way to show that the function is not linearly separable is to test for *k*-asummability. Since the only minterms for the output of 1 are at 100 and 111 we only need to test for 2-asummability. This means adding the two minterms together to get 211 and looking for any combination of minterms corresponding to 0s which also add up to 211. One example is the pair of minterms 101 and 110 which add up to 211. This shows that the function is not 2-asummable, so the function is therefore not linearly separable.

4 The following table shows the function with the appropriate weights. Also shown are the thresholds and the corresponding outputs. When all of the outputs are put through an exclusive-or, the function produced is the desired one.

x_1	x_2	x_3	*net*	$t_1=4$	$t_2=5$	$t_3=7$	y
0	0	0	0	0	0	0	0
0	0	1	1	0	0	0	0
0	1	0	2	0	0	0	0
0	1	1	3	0	0	0	0
1	0	0	4	1	0	0	1
1	0	1	5	1	1	0	0
1	1	0	6	1	1	0	0
1	1	1	7	1	1	1	1

CHAPTER 10

Implementation

CHAPTER OVERVIEW

This chapter gives describes some of the ways that have been tried to physically implement neural networks either electronically or optically. It explains:

- how neurons can be built electronically using analogue, digital or pulsed data
- how neurons can be built optically using switches or holograms

10.1 Introduction

Historically, neural networks have been built using all kinds of technology – mechanical, electrical and more recently optical. However, many of the neural networks that have been described in this book have never actually been implemented other than as simulations on conventional computers.

The study of neural networks is still a relatively new subject, and as such new ideas are still appearing. Inevitably, these ideas have to be shown to work before being taken up in any commercial sense, so implementation has lagged behind invention somewhat. Nevertheless, as the demand for more neurons increases, simulations become unwieldy and time-consuming, even on parallel computers. So the search for ways of constructing artificial neurons has started.

Networks that are currently being constructed fall into two categories – special-purpose architectures and general-purpose (Treleaven, 1988). The special purpose architectures directly implement a specific neural network such as the Hopfield network or the multi-layered perceptron. General-purpose architectures allow any network to be implemented, and usually involve some form of general-purpose parallel computer, with its associated programming language.

In the following sections some of the possible technologies available to construct neural networks are outlined and examples are given, where possible, of existing implementations. These include both electronic and optical implementations.

10.2 Electronic neural networks

In all implementations of neural networks, the following three operations are required:

- weighting the inputs
- summing the weighted inputs
- performing some non-linear function on the weighted sum

If the neural networks are to be implemented electronically, then the first decision that has to be made is whether to use analogue or digital representations.

10.2.1 Analogue electronics

Analogue electronic circuits provide a fast and relatively simple means of implementing neural networks. If inputs are voltages, simple resistances can be used to provide the weights. Ohm's law states:

$$V = IR$$

where V is a voltage, I is a current and R is a resistance. If we use conductance, G, where $G = 1/R$, then this becomes:

$$V = I/G \text{ or } I = GV$$

Thus, the current is the product of the input voltage and the weight is the conductance of a resistor. The weights can only be positive, so if negative weights are needed, the voltage input has to be inverted somehow.

If all of the weighted inputs are connected together, Kirchhoff's current law can be applied. This states that the sum of the currents into a junction is zero, or put another way, the current flowing out of a junction is equal to the sum of the currents flowing in. Figure 10.1 shows a simple passive network, constructed using resistors only.

For the network, the output voltage, V_o is:

$$V_o = R_o \times \sum V_i G_i$$

This circuit is simple, but would suffer from non-linear effects due to load resistances. An improvement can be made by using an operational-amplifier (op-amp), as shown in Figure 10.2.

The output voltage is now given as:

$$V_o = \left(1 + \frac{R_1}{R_2}\right) \sum V_i G_i$$

A non-linearity such as a hard-limiter can be provided using another op-amp. However, if we know that a hard-limiter is required as the output function then we can just use one op-amp without feedback. This also allows a negative input, which can be used to provide negative weights. The inputs are divided into two sets, V_i which are inputs with positive

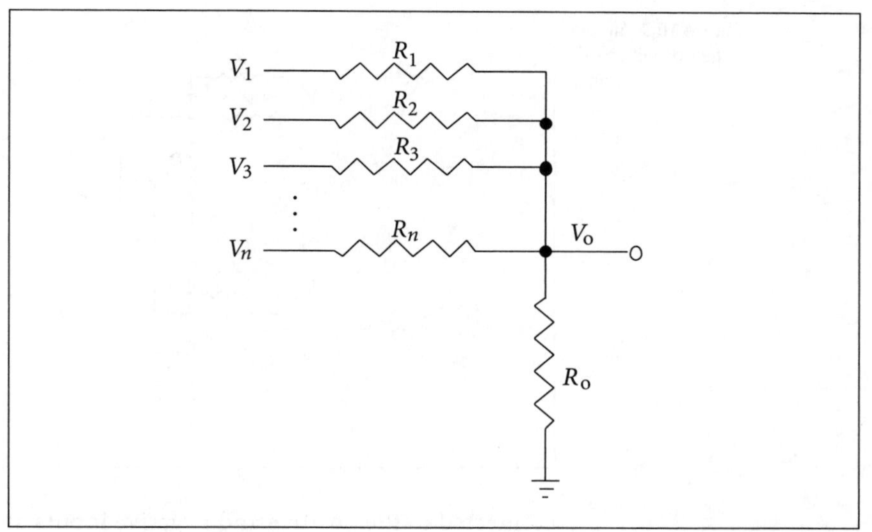

Figure 10.1 A passive network which implements a weighted sum

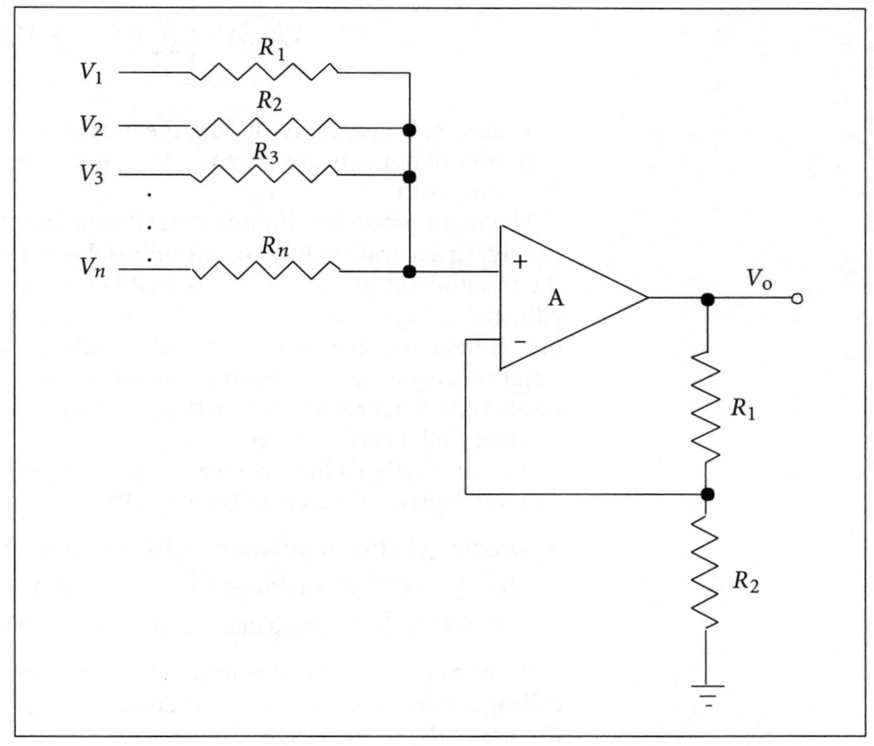

Figure 10.2 Op-amp circuit for the weighted sum

weights and V_j which are inputs associated with negative weights. A resistor network like that in Figure 10.1 is used for both sets, to find a positive weighted sum and the negative weighted sum which are then

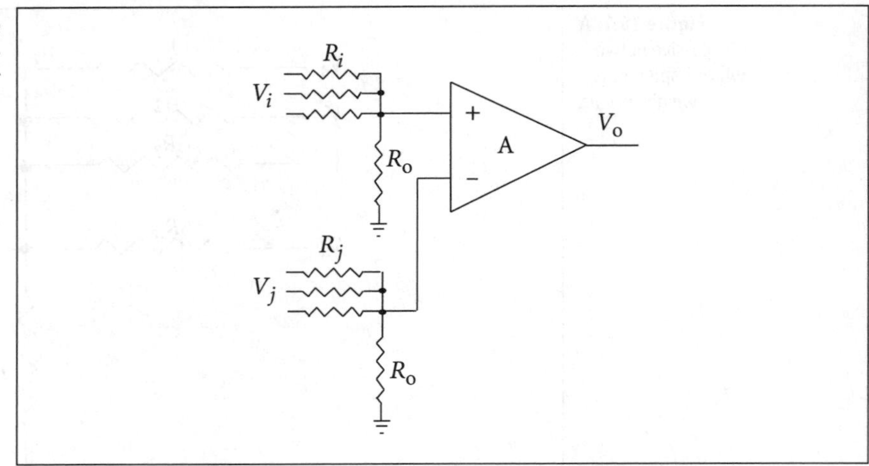

Figure 10.3 Simple neuron with hard-limiter

connected to the positive and negative inputs of the op-amp respectively. The circuit is shown in Figure 10.3, where the output is:

$$V_o = AR_o\left[\sum_i V_iG_i - \sum_j V_jG_j\right]$$

The gain, A, is very large, so that the output is $+V_{max}$ if the positive input is greater than the negative, and $-V_{max}$ if the negative input is greater than the positive.

The main problem with this type of analogue electronic circuit is providing accurate values for the adjustable resistances. In the ADALINE, the resistors on the inputs were variable potentiometers, and were adjusted using electric motors. Clearly, with the attempt to implement a neural network on a single VLSI chip, finding some way of providing the weights is needed, since resistors are large and variable resistors are impossible to fabricate. In CMOS technology, for example, once a resistor is fabricated its value is fixed.

Various methods have been attempted to get around this problem with varying degrees of success (Murray, 1989). These include:

- switching between different resistor values (Sivilotti *et al.*, 1985)
- digital weights in analogue circuits (Murray and Smith, 1988)
- dynamic weights using charge storage (Mackie *et al.*, 1988)

In the area of analogue neural networks, Carver Mead and his colleagues are probably the most dominant figures. It seems fruitless to discuss in depth the range of work that they have covered, and it is therefore suggested that for further investigation, one should read Carver Mead's book (Mead, 1989). In particular, the chapter on the silicon retina describes an analogue circuit that follows very similar principles to those that have just been discussed.

10.2.2 Digital electronics

The first type of digital electronic network is the RAM-based n-tuple network, such as the WISARD described in Chapter 4. There it was shown that the requirements of the system were calculated as the number of bits. For an P-bit input, the memory size needed is:

$$M = \frac{P \times 2^n \times D}{n}$$

where M is the total number of bits required, P is the number of bits in the input, D is the number of classes and n is the n-tuple size. For P and D with fixed value, the value of M varies as a function of n, as shown in the following table.

n	1	2	3	4	5	6	7	8	9	10	
M	2	2	2.67	4	6.4	10.7	18.3	32	56.9	102.4	$\times PD$

An alternative approach would be simply to implement the weighted sum as a digital operation using conventional logic gates and to provide a look-up table for the output non-linearity. Again, a number of methods have been implemented (Murray, 1989).

The weights could be stored as binary numbers. If the inputs are 0 and 1 then the multiplication is trivial - either add the weight if the corresponding input is 1 or not if the input is 0. Addition could be carried out using a cascaded full adder circuit, as shown in Figure 10.4.

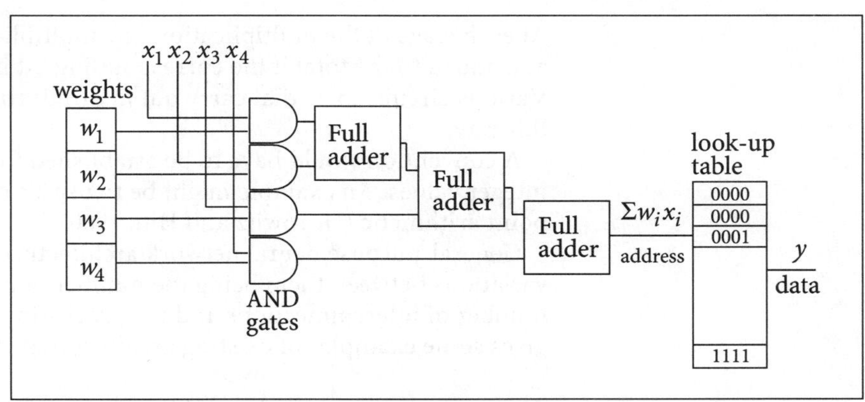

Figure 10.4 Digital neuron with look-up table

The truth table for a full-adder is:

A	B	C_p	C_n	S
0	0	0	0	0
0	0	1	0	1
0	1	0	0	1
0	1	1	1	0
1	0	0	0	1
1	0	1	1	0
1	1	0	1	0
1	1	1	1	1

If the inputs are anything other than binary, a multiplier would have to be constructed. Usually this is done using a 'shift-and-add' operation. For example, if 1011 is multiplied by 1001, the operation looks like this:

$$
\begin{array}{rrrrrrrr}
 & & & & 1 & 0 & 0 & 1 \\
 & & & \times & 1 & 0 & 1 & 1 \\
\hline
 & & & & 1 & 0 & 0 & 1 \\
 & & & 1 & 0 & 0 & 1 & \\
 & & 0 & 0 & 0 & 0 & & \\
 & 1 & 0 & 0 & 1 & & & \\
\hline
 & 1 & 1 & 0 & 0 & 0 & 1 & 1
\end{array}
$$

At each stage of the multiplication, the multiplicand is shifted to the left and added to the total if the corresponding bit in the multiplier is 1. Various circuits exist that carry out multiplication of binary numbers in this way.

A convention would have to be established for signed values and non-integer values. An example might be to use 2's complement and floating point arithmetic (Horowitz and Hill, 1980).

General-purpose neural network architectures are all digital, the variations between them being the number of processing elements, the number of interconnections and the processing speed. The following table gives some examples of existing architectures (Treleaven, 1988).

Neurocomputer	Number of PEs	Interconnects	Updates/s
HNC ANZA	30K	300K	25K
TRW MARK III	65K	1M	450K
TRW MARK IV	256K	4M	5M
IBM NEP	1M	4M	800K
NETSIM	256 × 27K	64K × 27K	4M

The ANZA Neurocomputer was developed by the Hecht-Nielsen Neurocomputer Corporation. It consists of a coprocessor board with the Motorola M68020 plus an M68881 floating point coprocessor with 4 Mbyte of dynamic RAM. It fits into the back of a PC, and comes with a library of software modules that allow a number of different neural network algorithms to be developed.

The MARK III and MARK IV were developed at TRW also by Robert Hecht-Nielsen. They are both add-on cards for the VME bus consisting of the M68020 and M68881 again.

The IBM NEP (Network Emulation Process) is a board that contains a T1320 signal processor together with high-speed interface chips and 68K words of SRAM for programs and data. It fits into the back of an IBM PC, and comes with its own software.

Finally, NETSIM was developed by Texas Instruments and Cambridge University. It is a collection of neural network simulator cards arranged in a three-dimensional array, each with its own 80188 processor, with a PC as the host computer. It supports many neural network algorithms, such as the multi-layer perceptron and the Hopfield network.

10.2.3 Pulsed data

In both of the previous implementations, the data is represented by voltage levels. An alternative approach is to use pulsed data where a stream of pulses to represent the data, with the information being represented by the frequency of the pulses. A number of arguments have been put forward for the advantages of this method (Murray, 1989) including:

- the 'multiply and add' function common to most neuron models can be carried out very efficiently using pulses
- it is possible to exploit the advantages of both analogue and digital circuitry
- there is a strong biological link, since actual neurons produce pulse streams

An example of a pulse-stream neuron (Murray and Smith, 1988) is one where the pulses are voltages with values of either 0 V or 5 V. A logic 0 is represented by no stream at all, so the frequency, f_0, is zero, and a logic 1 is represented by a stream of pulses at the maximum frequency, so $f_1 = f_{max}$. Values between 0 and 1 can then be represented by a stream of pulses with a frequency between 0 and f_{max}. This is shown in Figure 10.5, where the values of 0, 0.5 and 1 are shown as three pulse streams. The value of 0.5 is produced by a stream of pulses with a frequency of $0.5f_{max}$.

The pulses are produced by a switched oscillator, that is, an oscillator which produces a stream of pulses at its output but which can be switched on or off. The weights in such a system are stored as binary numbers, thus taking advantage of the precision inherent in digital values. The weights are scaled so that they are in the range –1 to +1. The value of each bit in a weight is linked to a rectangular wave generator. If the bit value is 1, the

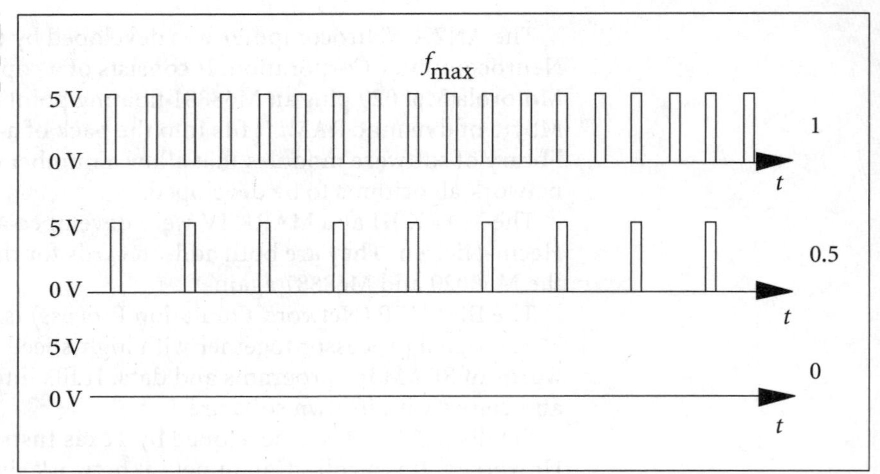

Figure 10.5 Logical value representation as a pulse stream

wave is generated and if the bit value is 0 the wave is not generated. The mark-space ratio of the rectangular wave is related to the relative position of the bit in the weight as shown in Figure 10.6.

The first bit in the weight represents the sign, '+' or '–'. If it is a 1 then it represents '–', and the signal is inhibitory. This bit is actually used to redirect the pulse stream so that the excitory and inhibitory signals are kept separate. The next bit represents the most significant bit, which, when it is 1, allows 50% of the pulse to stream through. This is done by taking the AND operation between the rectangular wave and the pulse stream. If the next bit is a 1 then a further 25% of the pulse stream gets through, and if the next bit is a 1 a further 12.5% and so on. In the example in Figure 10.6, 62.5% of the pulse stream gets through, which corresponds to a weight of 0.625. This would be produced if bit 1 is 0 to make the weight positive, bit 2 is 1, bit 3 is 0 and bit 4 is 1.

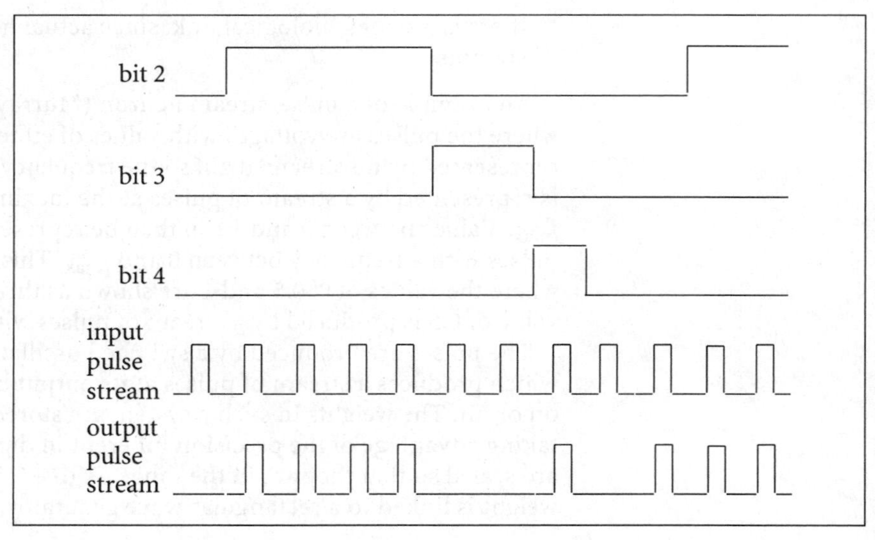

Figure 10.6 Rectangular waves and relative bit position for a weight of +0.625

When all of the pulse streams are weighted they are then summed and an output pulse stream produced if the sum is greater than zero. The way that the pulses are summed is to store charge in a capacitor. Each time the capacitor receives an excitory pulse, an amount of charge is added, and similarly, when it receives an inhibitory pulse, the same amount of charge is removed. The strength of each signal is represented by the number of pulses, so that after receiving the excitory and inhibitory signals, the amount of charge will be proportional to the number of excitory minus the number of inhibitory pulses.

The voltage on the capacitor is then passed on to a 'ring oscillator' such as the one shown in Figure 10.7. The output is therefore a pulse stream with a period determined by the delays of the inverters in the feedback loop.

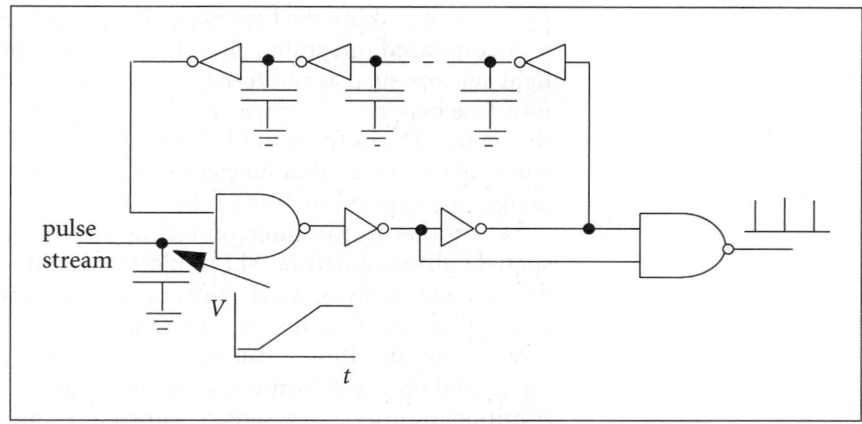

Figure 10.7 A ring counter

The advantages that are claimed for pulse stream neurons are that the circuits needed to implement the neurons are relatively simple and require only a few transistors. Thus many neurons can be fitted onto a small integrated circuit. Work is still in progress in this area, and better circuits, more specifically those with even fewer transistors, are being produced (Murray, 1989).

10.3 Optical neural networks

In recent years optical technology has been improving at a dramatic rate. In communications, optical fibres have become commonplace and optical switching is now possible. One advantage that optical technology brings, compared to conventional electronic communication, is that a larger bandwidth is possible resulting in faster transmission. Also a high degree of parallelism is possible because light waves are able to pass close by or even through each other in a linear medium without any problems arising. Even the simple lens can be thought of as being a massive parallel interconnector, allowing millions of points on an input image to be mapped to individual points on the output image.

A number of authors (Bell, 1986; Farhat, 1986; Psaltis, 1986; Caulfield *et al.*, 1989) have come to the conclusion that optical techniques are the best methods for implementing neural networks. Different neural networks have been implemented in this way, for example a Hopfield network (Farhat *et al.*, 1985), a bidirectional associative memory (Guest and TeKolste, 1987) and a multi-layered perceptron (Wagner and Psaltis, 1987).

Many of the properties and developments of optical systems are summarized in Feitelson's book (Feitelson, 1988) which is very readable and is therefore recommended for further information. In the following sections only the properties relevant to neural networks will be discussed.

10.3.1 Integrated opto-electronic systems

One way of making the best use of the properties of light is to arrange for a silicon-based integrated circuit to contain a two-dimensional array of light sensors such as photo-detecting diodes. This would provide an interface between an image and the electronic circuits required to process the image. The output could also be a two-dimensional array of light-emitting diodes, so that no electrical connections are necessary for the image-processing function of the chip.

Yet another way of interconnecting electronic and optical systems uses spatial light modulators (SLM). Electronic data is used to modulate a two-dimensional array of pixels, thus providing a means of entering data in parallel into an optical computer (Cathey *et al.*, 1989).

So, communication is one very useful property of optical systems. Integrated opto-electronic systems make use of this property, but still continue to process any signals using conventional electronic devices. The next step, therefore, is to find optical devices that can process the signals without the need to convert them to electronic signals.

10.3.2 Non-linear optical switches

Probably the most significant optical device that has been invented is the Fabry–Perot resonator, or etalon. Figure 10.8 shows a diagrammatic representation of a resonator.

The resonator consists of two parallel semi-transparent mirrors. Under normal operation, an incident beam would be reflected back and forth between the mirrors with some of the light getting transmitted. The light

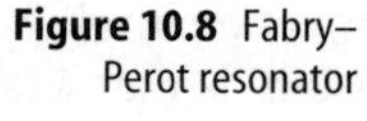
Figure 10.8 Fabry–Perot resonator

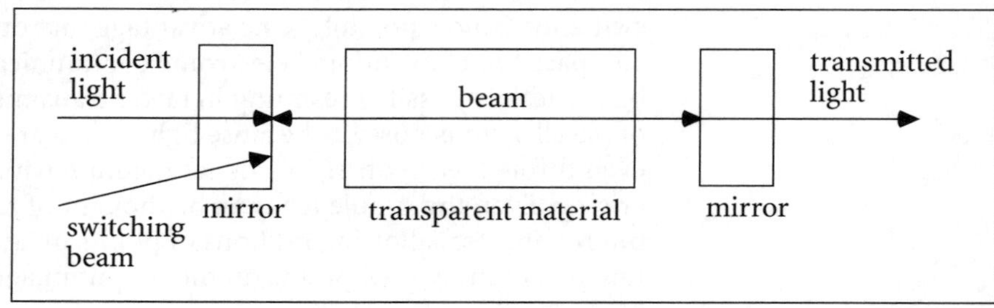

waves inside the resonator interfere, and at certain frequencies, the resonant frequencies, they interfere constructively producing high-intensity transmitted light.

Resonance depends on the wavelength of the light, λ, and the distance between the mirrors, d. For constructive interference the path difference for the various waves must be equal to a whole number of wavelengths. Thus:

$$2d = k\lambda$$

where k is an integer.

The resonator can be made into an optical switch by introducing a non-linear material between the mirrors. The material has the property called the photorefractive effect, where the refractive index of the material, n, changes with the intensity of the light. One way of producing such a device is to use a thin wafer of a non-linear crystal with the sides covered with a reflective layer. This device is called an etalon.

The refractive index of a medium is defined as the ratio of the speed of light in free space, c_0, to the speed of light in that medium, c_m.

$$n = \frac{c_0}{c_m}$$

The relationship between the frequency, f, and wavelength of light, λ, is:

$$c = f\lambda$$

Combining these two equations gives:

$$f = \frac{kc_0}{2dn}$$

So, as n increases, the resonant frequency decreases.

If the frequency of the incident beam is chosen to be just below the resonance frequency, the output is low intensity. Then, if a small additional light signal is applied, the intensity increases, so the refractive index increases and the resonance frequency comes down. As the resonance frequency comes down it will meet the frequency of the incident beam and the device will resonate, producing a high-intensity output. Thus an optical switch is produced, the output being either a small beam or a large beam depending on the small switching input being present or not.

Figure 10.9 shows the switching characteristics of a resonator. As the intensity of the switching beam increases, the intensity of the output remains low until a threshold is reached, at which point the intensity switches to some higher value. If the switching beam's intensity is now decreased, the intensity of the output stays high, then changes to a low value, but at a different point to the switch from low to high. This characteristic is known as hysteresis, and can be usefully employed in the construction of bistable devices.

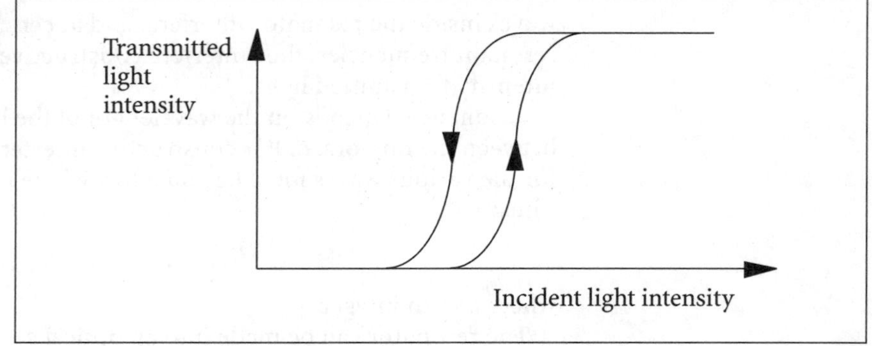

Figure 10.9 Switching characteristics

These switches, and similar arrangements, can be used to form a set of logic gates, such as AND, OR and NOT. The next step would be to use the digital approach to construct a neural network, but with optical rather than electronic switches. This improves speed and offers the opportunity for parallelism, but as yet has not been tried.

10.3.3 Holographic systems

It has been known for a relatively long time that holograms can perform as associative memories. Images are stored, and can be reproduced by shining the same reference beam that was used to record the image. It is also possible to reconstruct the original image using only a fraction of the original holographic material.

A planar hologram on a medium such as photographic film can direct any light beam on one side of it to any point on the other. If the film is 1 inch square, as many as 10 000 light sources can be connected to 10 000 light sensors (Abu-Mostafa and Psaltis, 1987). Clearly, this can be used as a massively parallel interconnector.

A volume hologram is made from a three-dimensional photorefractive crystal. The crystal contains electric charges which distribute themselves according to the intensity of light that is incident on the crystal. The refractive index is a function of local charge density, so the image projected on to the crystal is stored in terms of the value of the refractive index. The image information can be reconstructed simply by shining light on to the crystal.

These volume holograms, together with some optical switches can be combined to form an associative memory, capable of reconstructing an image even when it is presented with incomplete or shifted data. The arrangement is shown in Figure 10.10 (Abu-Mostafa and Psaltis, 1987; Hsu *et al.*, 1988).

The system consists of two volume holograms, each with the same stored images. Note that more than one image can be stored in the same hologram. It also contains a threshold device, which reflects light on one side according to what light is incident on the other side, a pinhole array and some lenses and mirrors.

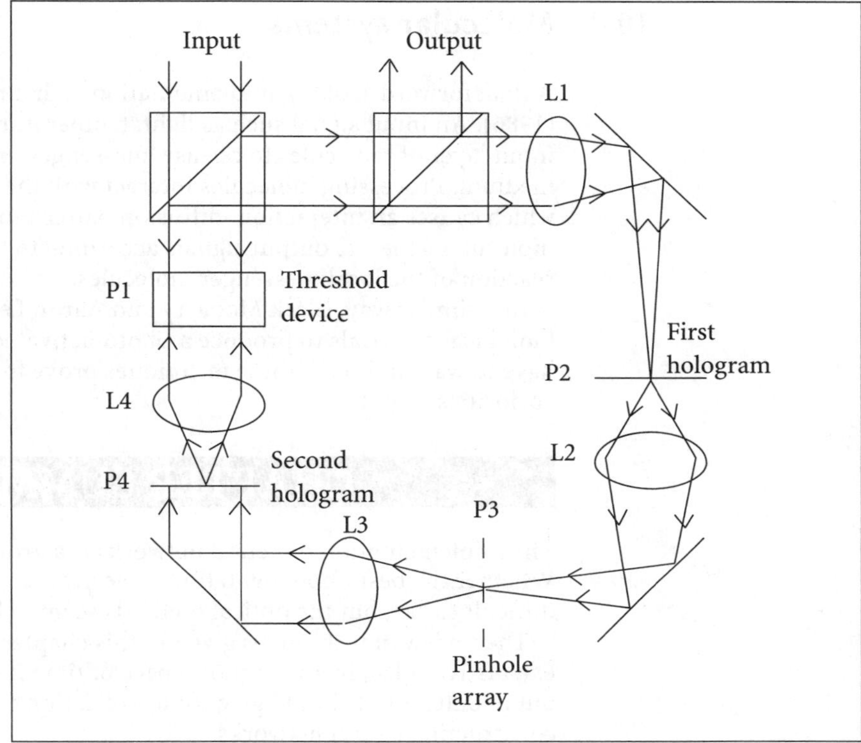

Figure 10.10 Pattern recognition system

First the network is trained. This is done by shining the images to be stored at the same time only spatially multiplexed onto the threshold device. The Fourier transform is taken by the lens L1 so that the magnitude spectrum of the images are stored holographically in P2 and then copied to P4. The image is reconstructed from P4 via lens L4 and a new Fourier transform of this image is stored in P2. The magnitude spectrum of an image is used because it is invariant to shifts in position, this information being found in the phase spectrum. An explanation of how a lens is used to find the Fourier transform of an image can be found in the literature (Feitelson, 1988).

When the system is operating, an image is presented which is transformed and correlated to the stored hologram in P2. A strong correlation produces a bright peak at one of the pinholes. This in turn causes the best matching image to be reconstructed from the hologram, P4. This shines on to the threshold device, which reflects some of the image back around the loop.

The result of this action is that the image stored in the memory that matches the input image is produced at the output. It does this even if the images do not match exactly, including the situations where the image is shifted slightly or where the image is incomplete.

10.4 Molecular systems

A final forward-looking implementation is that reported by Conrad (1986). An input signal such as light, temperature or pressure, causes an input layer of molecules to release 'messenger' molecules inside a medium. Processing molecules interact with the messenger molecules, which causes an interaction–diffusion pattern to be set up. Finally, output molecules generate output signals according to the outcome of the reaction of the local messenger molecules.

In a similar way, Clark Mobarry and Aaron Lewis (1986) have used biological materials to produce a photo-activated neural network. We will have to wait and see if these techniques prove to be just scientific curiosities or not.

CHAPTER SUMMARY

The implementation of neural networks is a growing field of study. Whether the best implementation is electronic, optical or chemical is difficult to say, but the optical methods seem to be the most promising.

The review of techniques given in this chapter is by no means exhaustive. Also, in a very short space of time it will probably look very out of date. But it should give some indication of the possible ways of constructing neural networks.

SELF-TEST QUESTIONS

1. In an analogue neural network, what value of resistance gives a conductance of 4.5?
2. How much memory is needed in a WISARD when the number of inputs is 64 and the number of discriminators is 16, if 2-tuples, 4-tuples or 8-tuples are used?
3. What is the product of the two binary numbers 1100 and 0110?
4. What does the 4-bit binary pulse stream look like for a weight of –0.75?

SELF-TEST ANSWERS

1. Conductance is the reciprocal of resistance. So, a conductance of 4.5 requires a resistance of 1 divided by 4.5 which is 0.222 ohms.
2. The table in Section 10.2.2 shows that the amount of memory is 2, 4 and 32 times *PD* respectively. If *P* is 64 and *D* is 16, then *PD* is 1024. So the amount of memory for 2-, 4- or 8-tuples is 2, 4 or 32 kbit, respectively.

3 The product is found using the shift and add convention.

```
     1100
×    0110
     0000
    1100
   1100
  0000    +
  1001000
```

4 A weight of −0.75 is represented in binary as 1110, where the first 1 represents the minus sign which directs the pulses to excitory or inhibitory inputs. The second 1 represents 0.5 and the third 1 represents 0.25. In Figure 10.11, pulses are allowed through when bit 2 and bit 3 are 1, but since bit 4 is 0 it blocks the pulse.

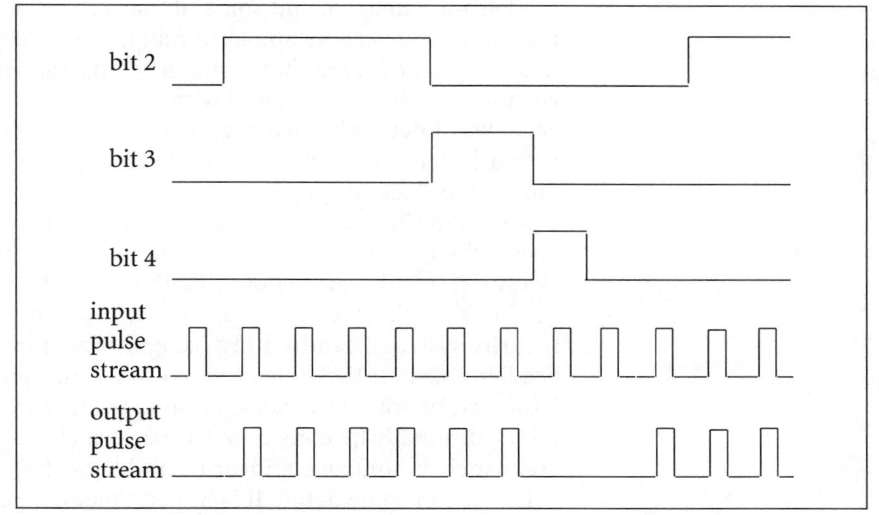

Figure 10.11 Pulses for the weight −0.75

Conclusions

This book started off by trying to answer what, at first, seemed like a very straightforward question – what is a neural network? What followed was an attempt at answering that question. The first chapter introduced some of the concepts that would be used throughout the book, such as the basic models for a neuron, and some of the architectures in which they are found. It soon became apparent that there is no simple answer, and that this is a research area that is still growing. Each new chapter introduced yet another architecture, and within that there would be several different varieties of network, each slightly different to the others. So many networks have been included, and yet there are still more that have been omitted for lack of space.

Again, in Chapter 1, the functional description of a neural network was given when it was said that a neural network is basically a pattern classifier. It is therefore not difficult to see why neural networks are being used in such diverse applications as face or fingerprint recognition in security systems, handwriting recognition, or even stock market prediction, since these clearly involve the recognition of patterns. However, neural networks are also appearing in applications which would not traditionally be considered as pattern classification. These include areas such as control engineering and logic design, which Chapters 8 and 9 describe in some detail. It is hoped that after reading these chapters, it is possible to see the way in which neural networks are being applied as pattern classifiers, but within a context of other disciplines.

Finally, Chapter 10 gives a very brief overview of the ways in which neural networks are being implemented. This aspect of neural networks is often overlooked, as software emulation is much more amenable. But until neural networks are actually implemented, the claim that they are fast due to their inherently parallel structure will be nothing more than a hollow statement.

Over the past 15 years, there has been an explosion in the number of people working with neural networks. Much of the theory, however, dates back to the previous two decades, and it has been surprising how little new theory has emerged with all this activity.

Much of the work that is currently going on in neural networks is in their application. What is emerging is that the successes are nearly always due to a combination of knowledge about the domain of the problem and knowledge about neural networks. In other words, it is no use applying neural networks to problems where little is already known in the hope that new understanding will emerge.

Nevertheless, we can expect to see new applications over the next 10 years, together with some theoretical developments, and no doubt we will start to see more products using neural networks. Certainly, one of the areas where there is a great deal of potential is in the integration of neural networks with other artificial intelligence methods, such as rule-based systems, and in particular, fuzzy rules (Kosko, 1992). Neural networks provide the adaptive capability in a system, but often it is easier to implement the resulting solution in another form, such as fuzzy logic. This also seems to be the only way that neural networks are going to find their way into safety-critical applications, where verification is all too important. Until neural networks can behave in predictable ways, they have to be governed by alternative methods such as rules (Picton *et al.*, 1991).

Finally, this book has been a basic introduction to neural networks. The alternative structures that can be formed from individual neurons include feedforward networks, feedback networks and self-organising networks. Within those structures there are alternative architectures and learning techniques. Rather than developing yet more networks or techniques to speed up learning, is it time to start looking at the network as the building block rather than the neuron? Research has already begun, and networks of networks (NoNs) are emerging. There is still a long way to go until we start to see truly intelligent machines, but we can be fairly confident that neural networks will be at the heart of any intelligent system from now on.

APPENDIX A

Derivation of the delta rule

ADALINE

Let the error function, E, be the mean squared error over all of the training patterns.

$$E = \frac{1}{P}\sum_{p=1}^{P} e_p$$

where P is the number of patterns in the training set.

$$e_p = \delta_p^2$$

$$\delta_p = (d_p - net_p)$$

$$net_p = \sum_{i=0}^{n} (w_i x_i)_p$$

$$E = \frac{1}{P}\sum_{p=1}^{P}\left(d_p - \sum_{i=0}^{n}(w_i x_i)_p\right)^2$$

The formula for the mean squared error is a quadratic expression in terms of all the weights. A quadratic function has only one minimum, so the mean squared error should only have one minimum with regard to each of the weights.

A fast way of finding a minimum is to use a gradient descent method, where the adjustments to the parameters (the weights in this case) are proportional to the derivative of the error function with respect to the weights, but opposite in sign.

$$\Delta w_i = -k \times \frac{\partial E}{\partial w_i}$$

where k is a constant.

Therefore, if the weights are to be adjusted using gradient descent, the derivative of the mean squared error function has to be found.

$$\frac{\partial E}{\partial w_i} = \frac{1}{P}\sum_{p=1}^{P}\frac{\partial e_p}{\partial w_i} = \frac{1}{P}\sum_{p=1}^{P}\frac{\partial e_p}{\partial \delta_p} \times \frac{\partial \delta_p}{\partial w_i}$$

$$\frac{\partial e_p}{\partial \delta_p} = 2\delta_p$$

$$\frac{\partial \delta_p}{\partial w_i} = \frac{\partial \delta_p}{\partial net_p} \cdot \frac{\partial net_p}{\partial w_i}$$

$$\frac{\partial \delta_p}{\partial net_p} = -1$$

$$\frac{\partial net_p}{\partial w_i} = x_{ip}$$

$$\frac{\partial E}{\partial w_i} = \frac{1}{P}\sum_{p=1}^{P}(2\delta_p \times -1 \times x_{ip}) = -\frac{2}{P}\sum_{p=1}^{P}\delta_p x_{ip}$$

$$\Delta w_i = \frac{2k}{P}\sum_{p=1}^{P}\delta_p x_{ip}$$

Therefore, adjusting the weights by an amount that is proportional to the mean of the values of $x\delta$ minimises the mean squared error. The delta rule adjusts the weights in proportion to the error times the input, and consequently should find the minimum by gradient descent. The multiplier $2k$ is usually replaced by the single Greek symbol eta, η.

Perceptrons

Single perceptron

First of all, consider a single perceptron with a sigmoid function instead of a hard-limiter. The learning rule is:

$$\Delta w_i = -k\frac{\partial E}{\partial w_i}$$

If the same error function as the original delta rule is used, that is the mean squared error over the training set, then:

$$E = \frac{1}{P}\sum_{p=1}^{P} e_p$$

$$e_p = (d_p - y_p)^2$$

$$y_p = \frac{1}{(1 + e^{-net_p})}$$

$$net_p = \sum_{i=0}^{n} (w_i x_i)_p$$

$$\frac{\partial E}{\partial w_i} = \frac{1}{P}\sum_{p=1}^{P} \frac{\partial e_p}{\partial w_i} = \frac{1}{P}\sum_{p=1}^{P} \frac{\partial e_p}{\partial y_p} \times \frac{\partial y_p}{\partial w_i}$$

$$= \frac{1}{P}\sum_{p=1}^{P} -2(d_p - y_p) \times \frac{\partial y_p}{\partial w_p}$$

$$= -\frac{2}{P}\sum_{p=1}^{P} (d_p - y_p) \times \frac{\partial y_p}{\partial net_p} \times \frac{\partial net_p}{\partial w_i}$$

$$= -\frac{2}{P}\sum_{p=1}^{P} (d_p - y_p) y_p (1 - y_p) x_{ip}$$

So

$$\Delta w_i = \frac{2k}{P}\sum_{p=1}^{P} x_{ip} y_p (1 - y_p)(d_p - y_p)$$

The term $y_p(1 - y_p)(d_p - y_p)$ is called δ_p, and the $2k$ is usually called the learning coefficient, η, so just as the original delta rule:

$$\Delta w_i = \frac{\eta}{P}\sum_{p=1}^{P} x_{ip}\delta_p$$

Multi-layered perceptron

Assume a three-layered network with a single output, n inputs and k units in the hidden layer. Also, assume that the weights are adjusted after every input pattern, so the error is the squared error and not the mean squared error.

The squared error at that output is $E = [d - x_3(1)]^2$

$$x_3(1) = f\left(\sum_{i=0}^{k} w_3(i,1) x_2(i)\right)$$

The value of the output of the qth unit in the hidden layer is:

$$x_2(q) = f\left(\sum_{i=0}^{n} w_2(i,q)x_1(i)\right)$$

For a particular weight in the qth hidden unit, say weight r:

$$\frac{\partial E}{\partial w_2(r,q)} = \frac{\partial E}{\partial x_3(1)} \times \frac{\partial x_3(1)}{\partial w_2(r,q)}$$

$$= \frac{\partial E}{\partial x_3(1)} \times \frac{\partial x_3(1)}{\partial x_2(q)} \times \frac{\partial x_2(q)}{\partial w_2(r,q)}$$

$$\frac{\partial E}{\partial x_3(1)} = -2[d - x_3(1)]$$

$$\frac{\partial x_3(1)}{\partial x_2(q)} = w_3(q,1)x_3(1)[1 - x_3(1)]$$

$$\frac{\partial x_2(q)}{\partial w_2(r,q)} = x_1(r)x_2(q)[1 - x_2(q)]$$

So:

$$\frac{\partial E}{\partial w_2(r,q)} = -\, x_1(r)x_2(q)[1 - x_2(q)]\; w_3(q,1)x_3(1)[1 - x_3(1)]\; [d - x_3(1)]$$

Using the same process of gradient descent as with the single perceptron, and substituting η for 2k, the change in weight is:

$$\Delta w_2(r,q) = \eta x_1(r)x_2(q)[1 - x_2(q)]\; w_3(q,1)x_3(1)[1 - x_3(1)]\; [d - x_3(1)]$$

or:

$$\Delta w_2(r,q) = \eta x_1(r)x_2(q)[1 - x_2(q)]\; w_3(q,1)\delta_3(1)$$

where

$$\delta_3(1) = x_3(1)[1 - x_3(1)][d - x_3(1)]$$

$$\Delta w_2(r,q) = \eta x_1(r)\delta_2(q)$$

where

$$\delta_2(q) = x_2(q)[1 - x_2(q)]\; w_3(q,1)\delta_3(1)$$

If there are more output neurons, the error E would be the sum:

$$E = [d_1 - x_3(1)]^2 + [d_2 - x_3(2)]^2 + \ldots\, [d_m - x_3(m)]^2$$

So, the derivative of the error with respect to the outputs is:

$$\sum_{i=1}^{m} \frac{\partial E}{\partial x_3(i)} = -2\sum_{i=1}^{m}[d_i - x_3(i)]$$

Consequently, the general form of the equation for δ for element q in the hidden layer of a three-layer network is:

$$\delta_2(q) = x_2(q)[1 - x_2(q)] \sum_{i=1}^{m} w_3(q,i)\delta_3(i)$$

For networks with more than one hidden layer the equation for δ in layer p is:

$$\delta_p(q) = x_p(q)[1 - x_p(q)] \sum_{i=1}^{m} w_{p+1}(q,i)\delta_{p+1}(i)$$

References

Abu-Mostafa, Y. S. and Psaltis, D. (1987) Optical neural computers, *Scientific American*, pp. 66–73.

Abu-Mostafa, Y. S. and St Jaques, J. M. (1985) Information capacity of the Hopfield model, *IEEE Transactions on Information Theory*, **IT-31**, 461–4.

Ackley, D. H., Hinton, G. E. and Sejnowski, T. J. (1985) A learning algorithm for Boltzmann machines, *Cognitive Science*, **9**, 147–69.

Aleksander, I. (1983) Emergent intelligent properties of progressively structured pattern recognition nets, *Pattern Recognition Letters*, **1**, 375–84.

Aleksander, I. (1989a) The logic of connectionist systems, in *Neural Computing Architectures*. North Oxford Academic.

Aleksander, I. (1989b) Canonical neural nets based on logic nodes, *Proc. 1st IEE International Conference on Artificial Neural Networks*, London, pp. 110–14.

Aleksander, I. and Burnett, P. (1984) The silicon neuron, in *Reinventing Man*. Penguin Books, Middlesex, Ch. 10.

Aleksander, I. and Stonham, T. J. (1979) Guide to pattern recognition using random-access memories, *IEE Proceedings Pt. E*, **2**(1), 29–40.

Aleksander, I. and Wilson, M. J. D. (1985) Adaptive windows for image processing, *IEE Proceedings Pt. E*, **132**(5), 233–45.

Aleksander, I., Stonham, T. J. and Wilkie, R. A. (1982) Computer vision systems for industry, *Digital Systems for Industrial Automation*, **1**(4), 305–20.

Aleksander, I., Thomas, W. V. and Bowden, P. A. (1984) WISARD: a radical step forward in image recognition, *Sensor Review*, pp. 120–4.

Amari, S.-I. and Maginu, K. (1988) Statistical neurodynamics of associative memory, *Neural Networks*, **1**, 63–74.

Anderson, C. W. (1989) Learning to control an inverted pendulum using neural networks, *IEEE Control Systems Magazine*, **9**(3), 31–7.

Austin, J. (1988) Grey scale *N*-tuple processing, *Proc. 4th International Conference on Pattern Recognition*, Cambridge, pp. 110–19.

Barto, A. G., Sutton, R. S. and Anderson, C. W. (1983) Neuronlike adaptive elements that can solve difficult learning control problems, *IEEE Transactions on Systems, Man, and Cybernetics*, **SMC-13**(5), 834–46.

Baum, E. B. (1986) Towards practical 'neural' computation for combinatorial optimisation problems, *Neural Networks for Computing*, AIP Conf. Proc. 151, Snowbird, Utah, pp. 47–52.

Bell, T. E. (1986) In search of an optical brain, *IEEE Spectrum*, pp. 49–51.

Bledsoe, W. W. and Browning, I. (1959) Pattern recognition and reading by machine, *Proc. Eastern Joint Computer Conference*, Boston, MA.

Broomhead, D. S. and Lowe, D. (1988) Multivariable functional interpolation and adaptive networks, *Complex Systems*, **2**, 321–55.

Brown, M. and Harris, C. (1994) *Neurofuzzy Adaptive Modelling and Control*. Englewood Cliffs, NJ: Prentice Hall.

Carpenter, G. A. (1989) Neural network models for pattern recognition and associative memory, *Neural Networks*, **2**(4), 243–57.

Carpenter, G. A. and Grossberg, S. (1987a) A massively parallel architecture for a self-organising neural pattern recognition machine, *Computer Vision, Graphics and Image Processing*, **37**, 54–115.

Carpenter, G. A. and Grossberg, S. (1987b) ART2: self-organisation of stable category recognition codes for analog input patterns, *Applied Optics*, **26**, 4919–30.

Carpenter, G. A. and Grossberg, S. (1988) The ART of adaptive pattern recognition by a self-organising neural network, *IEEE Computer*, 77–88.

Carpenter, G. A. and Grossberg, S. (1990) ART3: self-organisation of distributed pattern recognition codes in neural network hierarchies, *Proc. International Neural Network Conference*, Paris, France, pp. 801–4.

Cathey, W. T., Wagner, K. and Miceli, W. J. (1989) Digital computing with optics, *IEEE Proceedings*, **77**(10), 1558–72.

Caulfield, H. J., Kinser, J. and Rogers, S. K. (1989) Optical neural networks, *IEEE Proceedings*, **77**(10), 1573–83.

Conrad, M. (1986) The lure of molecular computing, *IEEE Spectrum*, **23**(10), 55–60.

Crisanti, A., Amit, D. J. and Gutfreund, H. (1986) Saturation level of the Hopfield model for neural network, *Europhysics Letters*, **2**(4), 337–41.

Davis, L. (1991) *Handbook of Genetic Algorithms*. New York: Van Nostrand Reinhold.

Dertouzos, M. L. (1965) *Threshold Logic: A Synthesis Approach*. Cambridge, MA: MIT Press.

Duda, R. O. and Hart, P. E. (1973) *Pattern Classification and Scene Analysis*. New York: Wiley.

Elgot, C. C. (1960) Truth functions realizable by single threshold organs, *Proc. Annual Symposium on Switching Circuit Theory and Logic Design*, pp. 225–45.

Elman, J. L. (1990) Finding structure in time, *Cognitive Science*, **14**, 179–211.

Farhat, N. H., Psaltis, D., Prata, A., Paek, E. (1985) Optical implementation of the Hopfield model, *Applied Optics*, **24**(10), 1469–75.

Farhat, N. H. (1986) Neural net models and optical computing; a brief overview, *SPIE Opt. and Hybrid Comp.* **634**, 307–11.

Feitelson, D. G. (1988) *Optical Computing*. Cambridge, MA: MIT Press.

Fukushima, K. (1975) Cognitron: a self-organising multilayered neural network, *Biological Cybernetics*, **20**, 121–36.

Fukushima, K. (1980) Neocognitron: a self-organising neural network model for a mechanism of pattern recognition unaffected by shift in position, *Biological Cybernetics*, **36**, 193–202.

Fukushima, K. (1989) Analysis of the process of visual pattern recognition by the neocognitron, *Neural Networks*, **2**(6), 413–20.

Fukushima, K. and Miyake, S. (1982) Neocognitron: a new algorithm for pattern recognition tolerant of deformations and shift in position, *Pattern Recognition*, **15**, 455–69.

Fukushima, K., Miyake, S. and Ito, T. (1983) Neocognitron: a neural network model for a mechanism of visual pattern recognition, *IEEE Transactions on Systems, Man, and Cybernetics*, **SMC-13**(5), 826–34.

Gabelman, I. J. (1962) The synthesis of Boolean functions using a single threshold element, *IRE Transactions*, **EC-11**, 639–42.

Giles, C. L. and Maxwell, T. (1987) Learning, invariance, and generalisation in higher-order neural networks, *Applied Optics*, **26**, 4972–8.

Gorse, D. and Taylor, J. G. (1989) Analysis of noisy RAM and neural nets, *Physica D*, **34**, 90–114.

Grant, P. M. and Sage, J. P. (1986) A comparison of neural network and matched filter processing for detecting lines in images, *Neural Networks for Computing*, AIP Conf. Proc. 151, Snowbird, Utah, pp. 194–9.

Grossberg, S. (1973) Contour enhancement, short-term memory, and constancies in reverberating neural networks, *Studies in Applied Mathematics*, **52**, 217–57.

Grossberg, S. (1976) Adaptive pattern classification and universal recoding, II: Feedback, expectation, olfaction, and illusions, *Biological Cybernetics*, **23**, 187–202.

Grossberg, S. (1987) *The Adaptive Brain*, Vols. I and II (Advances in Psychology, Vols. 42 and 43, ed. S. Grossberg). Amsterdam: North-Holland.

Grossberg, S. (1988) *Neural Networks and Natural Intelligence.* Cambridge, MA: MIT Press.

Guest, C. C. and TeKolste, R. (1987) Designs and devices for optical bidirectional associative memory, *Applied Optics*, **26**(23), 5055–60.

Haring, D. H. (1966) Multi-threshold elements, *IEEE Transactions on Electronic Computers*, **EC-15**, 45–65.

Haykin, S. (1999) *Neural Networks.* Englewood Cliffs, NJ: Prentice-Hall.

Hebb, D. O. (1949) *The Organisation of Behaviour.* New York: Wiley.

Hecht-Nielsen, R. (1987) Counter propagation networks, *Applied Optics*, **26**(23), 4979–84.

Hinton, G. E. (1985) Learning in parallel networks, *Byte*, 265–71.

Hopfield, J. J. (1982) Neural networks and physical systems with emergent collective computational abilities, *Proceedings of the National Academy of Science*, USA, Biophysics, 79, pp 2554-2558.

Hopfield, J. J. (1984) Neurons with graded responses have collective computational properties like those of two-state neurons, *Proceedings of the National Academy of Science, USA, Biophysics*, **81**, 3088–92.

Hopgood, A. A., Woodcock, N., Hallam, N. J. and Picton, P. D. (1993) Interpretating ultrasonic images using rules, algorithms and neural networks, *The European Journal of Non-Destructive Testing*, **2**(4), 135–49.

Hornick, K., Stinchcombe, M. and White, H. (1989) Multilayer feedforward networks are universal approximators, *Neural Networks*, **2**(5), 359–66.

Horowitz, P. and Hill, W. (1980) *The Art of Electronics.* Cambridge: Cambridge University Press.

Hsu, K., Brady, D. and Psaltis, D. (1988) Experimental demonstration of optical neural computers, in *Neural Information Processing Systems* (ed. Anderson, D. Z.). AIP, 377–86.

Hurst, S. L. (1978) *The Logical Processing of Digital Signals.* New York: Crane-Russak.

Ising, E. (1925) A contribution to the theory of ferromagnetism, *Z. Phys.*, **31**, 253.

James, W. (1990) Dumb intelligence, *Electronics World and Wireless World*, 194–9.

Kirkpatrick, S., Gellat, C. D. and Vecchi, M. D. (1983) Optimisation by simulated annealing, *Science*, **220**, 671–80.

Klír, J. and Valach, M. (1965) *Cybernetic Modelling.* London: Iliffe Books.

Kohonen, T. (1988) Associative memories and representations of knowledge as internal states in distributed systems, *Proc. European Seminar on Neural Computing*, London, UK, pp. 4/1–4/9.

Kohonen, T. (1984) *Self-Organisation and Associative Memory.* New York: Springer-Verlag.

Kosko, B. (1988) Bidirectional associative memories, *IEEE Transactions on Systems, Man and Cybernetics*, **SMC-18**, 42–60.

Kosko, B. (1987) Constructing an associative memory, *Byte*, 137–44.

Kosko, B. (1992) *Neural Networks and Fuzzy Systems: a Dynamical Systems Approach.* Englewood Cliffs, NJ: Prentice-Hall.

Lapedes, A. and Farber, R. (1988) How neural nets work, in *Neural Information Processing Systems* (ed. Anderson, D. Z.). AIP, 442–56.

Leonard, J. A. and Kramer, M. A. (1991) Radial basis function networks for classifying process faults, *IEEE Control Systems Magazine*, April 1991, pp. 31–8.

Leonard, J. A., Kramer, M. A. and Ungar, L. A. (1992) Using radial basis functions to approximate a function and its error bounds, *IEEE Transactions on Neural Networks*, **3**(4), 624–7.

Lippmann, R. P. (1987) An introduction to computing with neural nets, *IEEE ASSP Magazine*, 4-22.

Little, W. A. (1988) Spin models of neural networks – yesterday, today and tomorrow, in *Systems with Learning and Memory Abilities* (eds. Delacour, J. and Levy, J. C. S.). Amsterdam: Elsevier.

Mackie, W. S., Graf, H. P., Schwartz, D. B. and Denker, J. S. (1988) Microelectronic implementations of connectionist neural networks, in *Neural Information Processing Systems* (ed. Anderson, D. Z.). AIP, pp. 515–23.

McCulloch, W. S. and Pitts, W. (1943) A logical calculus of the ideas imminent in nervous activity, *Bulletin of Mathematical Biophysics*, **5**, 115–33.

McEliece, R., Posner, E., Rodemich, E. and Venkatesh, S. (1987) The capacity of the Hopfield associative memory, *IEEE Transactions on Information Theory*, **IT-33**(4), 461–82.

Mead, C. A. (1989) *Analog VLSI and Neural Systems*. Reading, MA: Addison-Wesley.

Meng, H. and Picton, P. D. (1992) Planning collision-free paths in time-varying environments, *Proc. 1st Int. Conf. Intelligent Systems Engineering*, Edinburgh, UK, pp. 310–15.

Meng, H. and Picton, P. D. (1993) A neural network motion predictor, *Proc. 3rd Int. Conf. on Artificial Neural Networks*, Brighton, pp. 177–81.

Michie, D. and Chambers, R. (1968) Boxes: an experiment in adaptive control, in *Machine Intelligence 2* (eds. Dale, E. and Michie, D.). Edinburgh: Edinburgh University Press, pp. 137–52.

Miller, W. T., Sutton, R. S. and Werbos, P. J. (1990) *Neural Networks for Control*. Cambridge, MA: MIT Press.

Minchinton, P. R., Bishop, J. M. and Mitchell, R. J. (1990) The Minchinton cell – analog input to the *n*-tuple net, Abstract in *Proc. International Neural Network Conference*, Paris, France, p. 599.

Minsky, M. and Papert, S. (1989) *Perceptrons*, 2nd edn. Cambridge, MA: MIT Press.

Mobarry, C. and Lewis, A. (1986) Implementation of neural networks using photoactivated conducting biological materials, *Proc. SPIE Int. Opt. Comp. Conf.*, **700**, 304–8.

Mow, C.-W.and Fu, K.-S. (1968) An approach for the realisation of multi-threshold threshold elements, *IEEE Transactions on Computers*, **C-17**, 32–46.

Muroga, S. (1971) *Threshold Logic and its Application*. New York: John Wiley Interscience.

Muroga, S., Tsuboi, T. and Baugh, C. R. (1970) Enumeration of threshold functions of eight variables, *IEEE Transactions on Computers*, **C–19**, 818–25.

Murray, A. F. (1989) Silicon implementations of neural networks, *Proc. 1st IEE Int. Conference on Artificial Neural Networks*, London, pp. 27–32.

Murray, A. F. and Smith, A. V. W. (1988) Asynchronous VLSI neural networks using pulse stream arithmetic, *IEEE Journal of Solid-State Circuits and Systems*, **23**(3), 688–97.

Pao, Y.-H. (1989) *Adaptive Pattern Recognition and Neural Networks*. Reading, MA: Addison-Wesley.

Parzen, E. (1962) On estimation of a probability density function and mode, *Annals of Mathematical Statistics*, **33**, 1065–76.

Pham, D. T. and Liu X. (1993) Identification of linear and non-linear dynamic systems using recurrent neural networks, *Artificial Intelligence in Engineering*, **8**, 67–75.

Picton, P. D. (1981) Realisation of multithreshold threshold logic networks using the Rademacher–Walsh transform, *IEE Proceedings Pt. E*, **128**(3), 107-113.

Picton, P. D. (1991a) Higher-order neural networks and the arithmetic transform, *Proc. 2nd Int. Conf. on Artificial Neural Networks*, Bournemouth, UK, 290–4.

Picton, P. D. (1991b) The relationship betwen Kohonen learning and Kalman filters, *Proc. IEE Colloquium on Adaptive Filtering, Non-Linear Dynamics and Neural Networks*, London, UK, 7/1–7/5.

Picton, P. D., Johnson, J. H. and Hallam, N. J. (1991) Neural networks in safety critical systems, *Proc. 3rd Int. Congress on Condition Monitoring and Diagnostic Engineering Management*, Southampton, UK, 17–24.

Picton, P. D. (1997) Neural and neuro-fuzzy control systems, *in Neural Network Analysis, Architectures and Applications* (ed. Browne, A.). Bristol: Institute of Physics Publishing, pp. 185–203.

Psaltis, D. (1986) Optical realisation of neural network models, *Proc. SPIE Int. Opt. Comp. Conf.*, **700**, 278–82.

Ritter, H., Martinez, T. and Schulten, K.(1992) *Neural Computation and Self-Organising Maps*. Reading, MA: Addison-Wesley.

Rosenblatt, F. (1958) The perceptron: a probabilistic model of information storage and organisation in the brain, *Psychological Review*, **65**, 386–408.

Rumelhart, D. E. and McClelland, J. L. (1986) *Parallel Distributed Processing.* Cambridge, MA: MIT Press.

Rumelhart, D. E. and McClelland, J. L. (1988) *Explorations in Parallel Distributed Processing.* Cambridge, MA: MIT Press.

Selviah, D. R., Midwinter, J. E., Rivers, A. W. and Lung, K. W. (1989) Correlating matched-filter model for analysis and optimisation of neural networks, *IEE Proceedings Pt. F*, **136**(3), 143–8.

Sherrington, D. (1989) Spin glasses and neural networks, in *New Developments in Neural Computing* (eds. Taylor, J. G. and Mannion, C. L. T.). Bristol: Adam Hilger, pp. 15–30.

Sivilotti, M. A., Emerling, M. E. and Mead, C. A. (1985) A novel associative memory implemented using collective computation, *Chapel Hill Conference on VLSI*, pp. 329–42.

Specht, D. F. (1990) Probabilistic neural networks, *Neural Networks*, **3**, 109–18.

Specht, D. F. (1991) A general regression neural network, *IEEE Trans. Neural Networks*, **2**, 568–76.

Steinbuch, K. (1961) Die lernmatrix, *Kybernetik*, **1**, 36–45.

Steinbuch, K. and Piske, U. A. (1963) Learning matrices and their application, *IEEE Transactions Electronic Computers*, **EC–12**, 846–62.

Szu, H. and Hartley, R. (1987) Fast simulated annealing, *Physics Letters A*, **122**(3,4), 157–62.

Treleaven, P. C. (1988) Parallel architectures for neurocomputers, *Proc. European Seminar on Neural Computing*, London, UK, pp. 6/1–6/27.

Ullmann, J. R. (1969) Experiments with the *n*-tuple method of pattern recognition, *IEEE Transactions on Computers*, **C–18**, 1135–7.

Van Hemmen, J. L. and Morgenstern, I. (eds.) (1986) *Heidelberg Colloquium on Glassy Dynamics*, Lecture Notes in Physics 275. Berlin: Springer-Verlag.

Von Neumann, J. (1956) Probabilistic logics and the synthesis of reliable organisms from unreliable components, in *Automata Studies* (eds. Shannon, C. E. and McCarthy, J.). Princeton: Princeton University Press, pp. 43–98.

Wagner, K. and Psaltis, D. (1987) Multilayer optical learning networks, *Applied Optics*, **26**(23), 5061–76.

Widrow, B. (1987) The original adaptive neural net broom-balancer, *Proc. Int. Symp. on Circuits and Systems*, pp. 351–7.

Widrow, B. and Hoff, M. E. (1960) Adaptive switching circuits, *1960 IRE WESCON Convention Record.* New York: IRE, pp. 96–104.

Widrow, B. and Smith, F. W. (1964) Pattern recognising control systems, in *Computer and Information Sciences* (eds. Tou, J. T. and Wilcox, R. H.). Washington, DC: Spartan Books, Cleaver Hume Press, pp. 288–317.

Widrow, B. and Winter, R. (1988) Neural nets for adaptive filtering and adaptive pattern recognition, *IEEE Computer*, 25–39.

Wieland, A. P. (1991) Evolving controls for unstable systems, in *Connectionist Models* (eds. Touretzky, D. S., Elman, J. L., Sejnowski, T. J. and Hinton, G. E.). San Mateo, CA: Morgan Kaufmann, pp. 91–102.

Winder, R. O. (1971) Chow parameters in threshold logic, *J. Ass. Com. Mach.*, **18**, 265–89.

Index